Mathematical Engineering

Series editors

Jörg Schröder, Essen, Germany
Bernhard Weigand, Stuttgart, Germany

Today, the development of high-tech systems is unthinkable without mathematical modeling and analysis of system behavior. As such, many fields in the modern engineering sciences (e.g. control engineering, communications engineering, mechanical engineering, and robotics) call for sophisticated mathematical methods in order to solve the tasks at hand.

The series Mathematical Engineering presents new or heretofore little-known methods to support engineers in finding suitable answers to their questions, presenting those methods in such manner as to make them ideally comprehensible and applicable in practice.

Therefore, the primary focus is—without neglecting mathematical accuracy—on comprehensibility and real-world applicability.

To submit a proposal or request further information, please use the PDF Proposal Form or contact directly: *Dr. Jan-Philip Schmidt, Publishing Editor (jan-philip. schmidt@springer.com)*.

More information about this series at http://www.springer.com/series/8445

Vissarion Papadopoulos · Dimitris G. Giovanis

Stochastic Finite Element Methods

An Introduction

 Springer

Vissarion Papadopoulos
School of Civil Engineering
National Technical University of Athens
Athens
Greece

Dimitris G. Giovanis
School of Civil Engineering
National Technical University of Athens
Athens
Greece

ISSN 2192-4732   ISSN 2192-4740 (electronic)
Mathematical Engineering
ISBN 978-3-319-87811-9   ISBN 978-3-319-64528-5 (eBook)
https://doi.org/10.1007/978-3-319-64528-5

Printed on acid-free paper

This Springer imprint is published by Springer Nature
The registered company is Springer International Publishing AG
The registered company address is: Gewerbestrasse 11, 6330 Cham, Switzerland

Preface

This book covers the basic topics of computational stochastic mechanics, while focusing on the stochastic analysis of structures in the framework of the finite element method (FEM). It is addressed to students of the postgraduate programme of the School of Civil Engineering at the National Technical University of Athens (**NTUA**). It is a self-contained book and aims at establishing a solid background on stochastic and reliability analysis of structural systems, such that it will enable future engineers to better manage the concepts of analysis and design in the presence of uncertainty as imposed in almost all modern engineering requirements.

Computational stochastic[1]mechanics is a field of mechanics that first appeared in the 70s in order to describe mechanical systems that show unpredictable behavior due to inherent uncertainty. Since then, it has evolved into a self-contained and prolific science field and has brought forth a wide range of sophisticated and well-established methodologies for the quantification of uncertainties inherent in engineering systems. Due to the rapidly growing availability of large-scale and low-cost computer power, the intensive computational demands of these techniques are becoming more and more manageable. This fact, has made the quantification of uncertainty increasingly popular in the engineering community; the progress achieved over the last decade in applying the theory of stochastic processes within the framework of classical engineering has led into higher levels of structural reliability with respect to traditional empirical safety factor approaches, both in terms of safety and economy. As a result, design engineers can now take rational and quantified risk mitigation measures to face the random nature of various parameters (material and geometric properties, loading conditions, etc).

Parameter uncertainty quantification and methods to predict uncertainty propagation on the response of structural systems have become an essential part of the analysis and design of engineering applications. *Stochastic analysis* and in particular the Stochastic Finite Element Method (SFEM), is a valuable and versatile tool for the

---

[1] The word "stochastic" as in "being random" derives from the Greek verb: *stochazomai*, "aim" which in one of its versions meant "trying to guess".

estimation of the response variability and the reliability of stochastic systems. A stochastic structural system is defined as a system with inherent uncertainties in its material, its geometric properties and its boundary conditions, subjected to stochastic (and/or deterministic) excitation. Implementation of SFEM leads to the estimation of the system's response variability which is a direct measure of the sensitivity of the system's response to the scatter of uncertain input parameters. It also leads to the evaluation of its reliability, an important factor in the design procedure which investigates the likelihood of the structure to fulfilling its design requirements. Response variability and reliability are closely tied to the concepts of performance-based engineering which is currently the way to apply design criteria to structures and structural components. These criteria are associated with the frequentist violation of various performance limit states, which are in turn linked to various engineering demand parameters such as stresses, displacements, etc.

There follows a brief outline of the book:

Chapter 1 introduces the fundamentals of the stochastic process theory and its applications. The definition of a stochastic process is given first, followed by a description of its characteristic functions and moments. Moreover, the meaning of ergodicity and stationarity are discussed and a description of the power spectrum is made.

Chapter 2 describes various methods used for the simulation of a stochastic process such as point discretization methods as well as the most popular Karhunen–Loève and spectral representation series expansion methods. Methods for the simulation of non-Gaussian fields are then presented followed by solved numerical examples. The first two chapters together with the appendixes, establish the necessary background for the rest of the book.

Chapter 3 contains the stochastic version of the virtual work approach which leads to the fundamental principles of the Stochastic Finite Element Method (SFEM). The resulting stochastic partial differential equations are solved with either nonintrusive simulation methods, such as the Monte Carlo simulation, or intrusive approaches such as the versatile spectral stochastic finite element method. Both approaches have proved to be powerful tools for the analysis of stochastic finite element systems and are described in detail along with additional approximate methodologies such as the Neumann and Taylor series expansion methods. Some exact analytic solutions that are available for statically determinate stochastic structures are next presented based on the so-called variability response functions and extended to general stochastic finite element systems. Illustrative examples are provided and discussed.

Chapter 4 is devoted to reliability analysis methods with emphasis given on those developed over the past two decades. The fundamental principles and basic reliability analysis methods are presented, namely the first- and second-order moments and the Monte Carlo simulation. For practical reliability problems, the latter require disproportionate computational effort. To overcome this liability, variance reduction-based simulation methods reduce the number of the Monte Carlo simulations required for an accurate prediction of the probability of failure; importance sampling, line sampling, and subset simulation are sufficiently deployed in this book. The use of artificial neural networks as effective surrogate models in the framework of reliability analysis is finally discussed. Examples of reliability analysis in real- world applications are presented illustrating the potential of each method and its relative advantages and disadvantages.

With this book, we wish to clarify in a scientific and concise way the admittedly difficult principles of the stochastic process theory and its structural engineering applications. We tried to simplify the theory in a way that the various methodologies and computational tools developed are nevertheless fully outlined to a new audience. To this purpose three appendixes were added that address the basic probability theory and random variables. In Appendix A, the basics of set theory are described followed by definitions of probability (classical, geometric, frequentist, conditional), in Appendix B the definition of a random variable is given together with the functions that are necessary for its description while in Appendix C the modified Metropolis–Hastings algorithms is presented and a MATLAB code is provided in order for the students to efficiently implement a reliability method called subset simulation. A number of unsolved problems are included at the end of each chapter.

Athens, Greece   Vissarion Papadopoulos
                                                                                 Dimitris G. Giovanis

Acknowledgements

Springer and the authors would like to acknowledge Ms. Eleni Filippas and Mr. Ioannis Kalogeris editing contribution and would like to thank them for their valuable comments and support which improved the clarity of the text.

Contents

1 Stochastic Processes 1
  1.1 Moments of Random Processes 3
      1.1.1 Autocorrelation and Autocovariance Function 4
      1.1.2 Stationary Stochastic Processes 5
      1.1.3 Ergodic Stochastic Processes 6
  1.2 Fourier Integrals and Transforms 7
      1.2.1 Power Spectral Density Function 8
      1.2.2 The Fourier Transform of the Autocorrelation
            Function 10
  1.3 Common Stochastic Processes 12
      1.3.1 Gaussian Processes 12
      1.3.2 Markov Processes 12
      1.3.3 Brownian Process 14
      1.3.4 Stationary White Noise 15
      1.3.5 Random Variable Case 17
      1.3.6 Narrow and Wideband Random Processes 17
      1.3.7 Kanai–Tajimi Power Spectrum 18
  1.4 Solved Numerical Examples 20
  1.5 Exercises 25

2 Representation of a Stochastic Process 27
  2.1 Point Discretization Methods 28
      2.1.1 Midpoint Method 29
      2.1.2 Integration Point Method 29
      2.1.3 Average Discretization Method 29
      2.1.4 Interpolation Method 30
  2.2 Series Expansion Methods 30
      2.2.1 The Karhunen–Loève Expansion 30
      2.2.2 Spectral Representation Method 35
      2.2.3 Simulation Formula for Stationary Stochastic Fields 39

2.3   Non-Gaussian Stochastic Processes 41
2.4   Solved Numerical Examples 42
2.5   Exercises 44

3   Stochastic Finite Element Method 47
3.1   Stochastic Principle of Virtual Work 47
3.2   Nonintrusive Monte Carlo Simulation 49
      3.2.1   Neumann Series Expansion Method 50
      3.2.2   The Weighted Integral Method 51
3.3   Perturbation-Taylor Series Expansion Method 52
3.4   Intrusive Spectral Stochastic Finite Element Method (SSFEM) ... 54
      3.4.1   Homogeneous Chaos 54
      3.4.2   Galerkin Minimization 56
3.5   Closed Forms and Analytical Solutions with Variability
      Response Functions (VRFs) 59
      3.5.1   Exact VRF for Statically Determinate Beams 60
      3.5.2   VRF Approximation for General Stochastic FEM
              Systems 63
      3.5.3   Fast Monte Carlo Simulation 64
      3.5.4   Extension to Two-Dimensional FEM Problems 65
3.6   Solved Numerical Examples 66
3.7   Exercises 68

4   Reliability Analysis 71
4.1   Definition 72
      4.1.1   Linear Limit-State Functions 73
      4.1.2   Nonlinear Limit-State Functions 75
      4.1.3   First- and Second-Order Approximation Methods 77
4.2   Monte Carlo Simulation (MCS) 78
      4.2.1   The Law of Large Numbers 78
      4.2.2   Random Number Generators 79
      4.2.3   Crude Monte Carlo Simulation 80
4.3   Variance Reduction Methods 82
      4.3.1   Importance Sampling 83
      4.3.2   Latin Hypercube Sampling (LHS) 83
4.4   Monte Carlo Methods in Reliability Analysis 84
      4.4.1   Crude Monte Carlo Simulation 84
      4.4.2   Importance Sampling 85
      4.4.3   The Subset Simulation (SS) 85
4.5   Artificial Neural Networks (ANN) 88
      4.5.1   Structure of an Artificial Neuron 88
      4.5.2   Architecture of Neural Networks 90
      4.5.3   Training of Neural Networks 91
      4.5.4   ANN in the Framework of Reliability Analysis 94

4.6   Numerical Examples   96
4.7   Exercises   98

Appendix A: Probability Theory   99

Appendix B: Random Variables   109

Appendix C: Subset Simulation Aspects   131

References   135

List of Figures

Fig. 1.1 A stochastic function is a mapping from a random event to the space of functions 2

Fig. 1.2 Realizations of **a** the Young's modulus of a beam and **b** a gambler's gain 2

Fig. 1.3 Four realizations of a stochastic process $X_i(t)$ with $i = 1, 2, 3, 4$. At time t_n, with $n = 1, 2, \ldots$, the random process is a family of random variables $X_i(t_n)$ 3

Fig. 1.4 Forms of stationarity and their relevance 8

Fig. 1.5 Two power spectrums $S_1(\omega)$ and $S_2(\omega)$ corresponding to Eqs. (1.30) and (1.31), respectively, for $b = \sigma_X = 1$ 10

Fig. 1.6 State diagram of a Markov chain 13

Fig. 1.7 A typical random walk on the number line 14

Fig. 1.8 A typical walk in a Brownian motion 15

Fig. 1.9 **a** The autocorrelation function, **b** the corresponding power spectrum, and **c** sample realization of a typical stationary white noise 16

Fig. 1.10 Autocorrelation function and corresponding power spectrum of a stationary band-limited white noise process 17

Fig. 1.11 **a** The autocorrelation function, **b** the corresponding power spectrum, and **c** sample realization of a random variable 18

Fig. 1.12 Power spectrum of a **a** wideband stochastic process, **b** narrowband stochastic process 18

Fig. 1.13 The Kanai–Tajimi power spectrum 19

Fig. 1.14 A simply supported beam under a stochastically distributed load 25

Fig. 1.15 Power spectrum of a stochastic process 26

Fig. 2.1 Decaying eigenvalues from the solution of the Fredholm integral of the second kind for $M = 10$ 32

Fig. 3.1 Three-dimensional body and a representative three-dimensional finite element 48

Fig. 3.2 Sparsity pattern of **K** for the Gaussian case ($M = 2$, $p = 2$) 58

Fig. 3.3   Sparsity pattern of **K** for the Gaussian case ($M = 6$, $p = 4$)   59
Fig. 3.4   Configuration of statically determinate beam   61
Fig. 3.5   Representation of the stochastic field using the midpoint
              method and **a** $N = 1$ element, **b** $N = 2$ elements, **c** $N = 4$
              and **d** convergence of the variance of tip displacement as
              a function of the number of finite elements.   67
Fig. 3.6   VRF of statistically determinate beam   68
Fig. 3.7   Geometry, loading, finite element mesh, and material
              properties of the cylindrical panel. Note that although
              not shown, every little "rectangle" in the above figure
              is subdivided into two "triangular" finite elements   69
Fig. 3.8   Variability response function calculated using the 2D
              FEM-FMCS approach for the cylindrical panel shown
              in Fig. 3.7 and $\sigma_f = 0.4$ 69
Fig. 3.9   Variance of response displacement u_{A_v}: comparison
              of results using Eq. (3.77) and from brute force Monte
              Carlo simulations (MCS). Plots correspond to SDF$_3$
              with $b_x = b_y = 2.0$. The corresponding PDF
              is a truncated Gaussian   70
Fig. 3.10   Clumped rod with stochastic properties   70
Fig. 4.1   Schematic representation of the probability, the resistance
              R to be smaller than the value x of the action S   73
Fig. 4.2   Geometric interpretation of the safety index β - linear
              failure function   76
Fig. 4.3   Geometric interpretation of the safety index β - nonlinear
              failure function   76
Fig. 4.4   Schematic representation of the hit and miss method   81
Fig. 4.5   In LHS method, the samples are randomly generated
              sampling once from each bin   84
Fig. 4.6   Division of $G(\mathbf{X}) = G(X_1, X_2)$ into M subsequent
              subsets G_i   86
Fig. 4.7   Structure of a biological neuron   89
Fig. 4.8   Basic structure of an artificial neuron   89
Fig. 4.9   Two simple feed-forward ANNs, **a**—A multilayer network,
              and **b**—a single-layer network. The *square nodes* represent
              the input nodes while the *circular nodes* represent the basic
              processing neurons   91
Fig. 4.10   Basic idea of the backpropagation algorithm
              for a univariate error function $E_t(w)$ 93
Fig. 4.11   Training performance of an ANN over a set of data.   94
Fig. 4.12   Five-storey plane frame with data; loading, mode
              of failure, and load-displacement curve   96

Fig. 4.13   Performance of NN configuration using different number
             of hidden units   96
Fig. A.1   The sample space of throwing dice.   100
Fig. A.2   A Venn diagram is a tool used to show intersections,
             unions, and complements of sets.   102
Fig. A.3   Buffon's needle illustration   104
Fig. A.4   Partition of a sample space Ω in a countable collection
             of events.   107
Fig. B.1   Definition of a random variable as a mapping
             from $\Theta \rightarrow \mathbb{R}$.   109
Fig. B.2   Cumulative distribution function of a random variable   110
Fig. B.3   The probability $P(a < x \leq b)$ is estimated
             as the shaded area of the pdf between a and b   111
Fig. B.4   **a** A unimodal and **b** a multimodal (bimodal)
             distribution function   111
Fig. B.5   The pdf and cdf functions of a random variable
             with uniform distribution   112
Fig. B.6   Graphs of distributions with negative, zero
             and positive skewness.   113
Fig. B.7   Graph of distributions with negative, zero
             and positive kurtosis.   114
Fig. B.8   Estimation of the pdf for an increasing function
             of a random variable.   115
Fig. B.9   Graph of the function of the two random variables.   116
Fig. B.10   Perceptiles of the normal distribution for values
             less than one, two and three standard deviation
             away from the mean   119

List of Tables

Table 3.1   One-dimensional polynomial chaoses   56
Table 3.2   Two-dimensional polynomial chaoses   56
Table 4.1   Characteristics of random variables   96
Table 4.2   Characteristics of random variables for MCS-IS   96
Table 4.3   "Exact" and predicted values of P_f   97
Table 4.4   Required CPU time for the prediction of
             P_f. (☆ corresponds to 10000 MCS)   97

Chapter 1
Stochastic Processes

Generalizing the definition of a random variable (see Appendix B) to the space of functions, we can define a stochastic process X as a mapping (see Fig. 1.1) from the probability space $\{\Theta, \mathscr{F}, P\}$, Θ being the sample space, i.e., the space of all possible events, $\mathscr{F}$ a σ-algebra (see Sect. A.2.4) on Θ and P the probability measure, to a function of time (stochastic process) and/or space (stochastic field). In the general case, a stochastic field $X(t)$ with $t \in \mathbb{R}^n$ is n-dimensional (nD). Another characterization can be made based on whether X is described either by a single random variable (univariate, 1V) or by a random vector (multivariate, mV).

The notation that is used in the remaining of the book for a stochastic process is as follows:

$$1\mathrm{D} - 1\mathrm{V} : X(t, \theta)$$
$$n\mathrm{D} - 1\mathrm{V} : X(\boldsymbol{t}, \theta)$$
$$n\mathrm{D} - m\mathrm{V} : \boldsymbol{X}(\boldsymbol{t}, \theta).$$

In the notation of a stochastic process or field over θ, its dependence on θ is omitted for the remaining of this chapter. The stochastic function can be either discrete or continuous over time/space and, for every realization θ_j, $j = 1, \ldots, k$ is called a sample function. Therefore, a stochastic function $X(t)$ is an ensemble of $\{X_j(t) = X(t, \theta_j)\}$, $j = 1, \ldots, k$, continuous functions generated from k realizations of the parameter space (Fig. 1.3). Some common examples of stochastic processes are

- The modulus of elasticity $\mathrm{E}(x, \theta)$, along the length x of a beam (Fig. 1.2a)
- A gambler's gain $\mathrm{Gain}(t, \theta)$ as a function of time t, in a game of chance (Fig. 1.2b).

For example, if $\xi(\theta)$ is a random variable uniformly distributed in the interval $[0, 2\pi]$, then a cosinusoidal 1D-1V random process can be defined as the function $X(t, \theta) = \xi(\theta) \cdot \cos(t)$ of the time parameter t. At any given time t_i, $X(t_i, \theta)$ is a random

© Springer International Publishing AG 2018
V. Papadopoulos and D.G. Giovanis, *Stochastic Finite Element Methods*,
Mathematical Engineering, https://doi.org/10.1007/978-3-319-64528-5_1

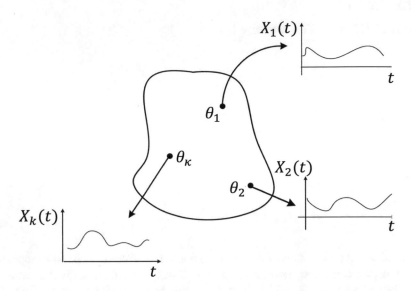

Fig. 1.1 A stochastic function is a mapping from a random event to the space of functions

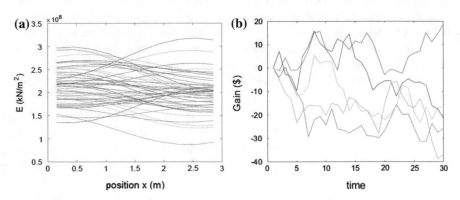

Fig. 1.2 Realizations of **a** the Young's modulus of a beam and **b** a gambler's gain

variable. As the stochastic process shifts over time instances $t_1, \ldots, t_n, t_{n+1}, \ldots$, it generates a sequence of random variables $\{X(t_1), \ldots, X(t_n), X(t_{n+1}), \ldots\}$ (Fig. 1.3).

Following the above-mentioned concept, an 1D-1V random process $X(t)$ is interpreted as a sequence of jointly distributed random variables $X(t_i)$ with $i = 1, \ldots, n$ for which we need to define their joint distribution. The n-th order joint cumulative distribution is defined as

$$F_{X(t_1), X(t_2), \ldots, X(t_n)}(x_1, x_2, \ldots, x_n) = P\big[X(t_1) \leq x_1 \ldots, X(t_n) \leq x_n\big], \qquad (1.1)$$

while the joint probability density function (pdf) of $X(t)$ is defined as

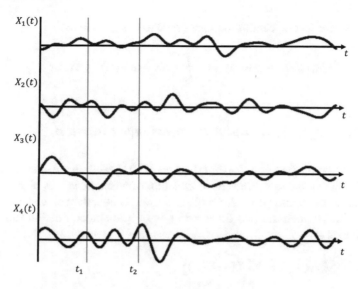

Fig. 1.3 Four realizations of a stochastic process $X_i(t)$ with $i = 1, 2, 3, 4$. At time t_n, with $n = 1, 2, \ldots$, the random process is a family of random variables $X_i(t_n)$

$$f_{X(t_1), X(t_2), \ldots, X(t_n)}(x_1 \ldots, x_n) = \frac{\partial^n}{\partial x_1 \ldots \partial x_n} F_{X(t_1), X(t_2), \ldots, X(t_n)}(x_1, \ldots, x_n) \quad (1.2)$$

1.1 Moments of Random Processes

Let $f_X(x)$ be the pdf of random variable $X(t)$ that represents the set of samples across the ensemble at time t. We can define the nth moment as

$$m_n(t) = \mathbb{E}[(X^n(t)] = \int_{-\infty}^{+\infty} x^n f_X(x) \mathrm{d}x \quad (1.3)$$

The first moment (mean value) $m_1(t) = \mu_X(t)$ is defined as

$$\mu_X(t) = \mathbb{E}[X(t)] = \int_{-\infty}^{+\infty} x f_X(x) \mathrm{d}x, \quad (1.4)$$

while the nth central moment is given by the following formula:

$$\mathbb{E}[(X(t) - \mu_X(t))^n] = \int_{-\infty}^{+\infty} (x - \mu_X(t))^n f_X(x) \mathrm{d}x \quad (1.5)$$

For example, the second central moment (variance) is defined as

$$\mathbb{E}\left[(X(t) - \mu_X(t))^2\right] = \int_{-\infty}^{+\infty} (x - \mu_X(t))^2 f_X(x)\mathrm{d}x \qquad (1.6)$$

1.1.1 Autocorrelation and Autocovariance Function

The autocorrelation function $R_X(t_i, t_j)$ of an 1D-1V random process $X(t)$ is an important function that quantifies the correlation between the values of $X(t)$ at two different instances t_i and t_j, for $i, j = 1, \ldots, n$ and is in certain cases a function of the difference between the two times $\tau = t_i - t_j$. The autocorrelation function of $X(t)$ is defined as

$$R_X(t_i, t_j) = \mathbb{E}\left[X(t_i)X(t_j)\right]$$
$$= \int_{-\infty}^{+\infty} \int_{-\infty}^{+\infty} x_i x_j f_{X(t_i), X(t_j)}(x_i, x_j)\mathrm{d}x_i\mathrm{d}x_j, \qquad (1.7)$$

where $X(t_i)$ and $X(t_j)$ are the values of the stochastic process at time instances t_i and t_j (i.e., two random variables) with joint pdf $f_{X(t_i), X(t_j)}(x_i, x_j)$. The autocorrelation function satisfies the following properties:

- **Symmetry**:
$$R_X(t_i, t_j) = R_X(t_j, t_i) \qquad (1.8)$$

- **Cauchy–Schwarz inequality**[1]:
$$R_X(t_i, t_j)^2 \leq R_X(t_i, t_i)R_X(t_j, t_j) \qquad (1.9)$$

- **Nonnegative definite**:

$$\sum_{-\infty}^{+\infty} \sum_{-\infty}^{+\infty} R_X(t_j - t_k)g(t_j)g(t_k) \geq 0$$

$$\lim_{\tau \to 0} R_X(\tau) = \mathbb{E}[X(t)^2] \qquad (1.10)$$
$$\lim_{\tau \to \infty} R_X(\tau) = 0$$

for all functions $g : \mathbb{Z} \longmapsto \mathbb{C}$ for which the above summation converges. In a similar manner, the autocovariance function can be defined relatively to the autocorrelation function as

---

[1]The Cauchy–Schwarz inequality was named after Augustin-Louis Cauchy (1789–1857) and the German mathematician Hermann Amandus Schwarz (1843–1921).

$$C_X(t_i, t_j) = R_X(t_i, t_j) - \mu_X(t_i)\mu_X(t_j) \qquad (1.11)$$

For a zero-mean stochastic process ($\mu_X(t_i) = \mu_X(t_j) = 0$), the covariance is equal to the autocorrelation function. The mean and autocorrelation functions of a stochastic process provide a partial characterization of the process, referred to as second-moment characterization. It is clear that stochastic processes with the same second-moment characteristics can have very different sample properties:

$$C_X(t_i, t_j) = R_X(t_i, t_j) \qquad (1.12)$$

The value of $C_X(t, t)$ on the diagonal $t_i = t_j = t$ is equal to the variance of the stochastic process at time t. Thus, the variance of the stochastic process can be defined as

$$\text{Var}(X(t)) = C_X(t, t) = \mathbb{E}\big[(X(t) - \mu_X(t))^2\big] \qquad (1.13)$$

Consequently, if we normalize the autocorrelation function by subtracting the mean and dividing by the variance we obtain the autocorrelation coefficient

$$\rho_X(t_i, t_j) = \frac{C_X(t_i, t_j)}{\sqrt{C_X(t_i, t_i)}\sqrt{C_X(t_j, t_j)}} \quad \text{with} \quad |\rho_X(t_i, t_j)| \leq 1 \qquad (1.14)$$

In case we want to quantify the dependence between two different stochastic processes $X(t), Y(t)$, we use the so-called cross-correlation function $R_{XY}(t_i, t_j)$, which is a measure of the lag of the one relative to the other, as

$$R_{XY}(t_i, t_j) = \mathbb{E}\big[X(t_i)Y(t_j)\big] = \int_{-\infty}^{+\infty} \int_{-\infty}^{+\infty} x_i y_j f_{X(t_i), X(t_j)}(x_i, y_j) \mathrm{d}x_i \mathrm{d}y_j \qquad (1.15)$$

The properties of the autocorrelation also stand for the cross-correlation function. In a similar manner to the autocovariance function, the cross-covariance function $C_{XY}(t_i, t_j)$ is defined if we subtract the product mean value from the cross-correlation function

$$C_{XY}(t_i, t_j) = R_{XY}(t_i, t_j) - \mu_X(t_i)\mu_Y(t_j) \qquad (1.16)$$

1.1.2 Stationary Stochastic Processes

An important category of stochastic processes is the ones whose complete probabilistic structure is invariant to a shift in the parametric origin:

$$f_X(x_1, t_1; \ldots, x_k, t_k) = f_X(x_1, t_1 + a; \ldots, x_k, t_k + a) \qquad (1.17)$$

These stochastic processes are called stationary (or homogeneous in the case of random fields). A result of stationarity is that the mean and variance (as well as higher moments) do not change when shifted in time

$$\mu_X(t) = \mu_X(t + a) = \mu_X, \quad \text{Var}(X(t)) \rightarrow \text{Var}(X) \tag{1.18}$$

An important consequence of this time (space) invariance is that the autocorrelation function at times t_i, t_j is independent at the time instants t_i, t_j and depends only on the relative distance $\tau = t_j - t_i$. This leads to the following expression and bounds for R_X and C_X:

$$R_X(t_i, t_j) = R_X(t_i + a, t_j + a) = R_X(\tau), \quad C_X(t_i + a, t_j + a) = C_X(t_i, t_j) = C_X(\tau)$$
$$|R_X(\tau)| \leq R_X(0) = \mathbb{E}[X(t)^2], \qquad |C_X(\tau)| \leq \sigma_X^2 = \text{Variance} \tag{1.19}$$

There are two categories regarding the form of stationarity of a random process:

- **Wide-sense stationary (WSS)**: The mean value is constant over time and the autocorrelation is a function of the time lag τ
- **Strictly stationary (SS)**: All moments of the stochastic process are constant over time.

1.1.3   Ergodic Stochastic Processes

In order to apply the theory of random processes, robust estimates of the mean and autocorrelation function are required, based on measurements. This estimation is done by using a sufficient large number of sample realizations of the random process.[2] From the total of sample realizations, the ensemble average at each time is defined as the average of the entire population of sample realizations. For n realizations of a sample functions $X(t)$, the ensemble average at time t is defined as

$$\mathbb{E}[X(t)] = \frac{X_1(t) + X_2(t) + \cdots + X_n(t)}{n} \tag{1.20}$$

In practice, one can have only a very small number of sample functions of the random process and/or a long, single observation of one sample function. So, the question is if statistical averages of the stochastic process can be determined from a single sample function. The answer to this is yes and it comes from the definition of ergodicity which stands only for stationary random processes.

---

[2]The study of ergodic theory was first introduced in classical statistical mechanics and kinetic theory in order to relate the average properties of a system of molecules to the ensemble behavior of all molecules at any given time. The term ergodic was first introduced by Ludwig Boltzmann (1844–1906).

A stationary random process $X(t)$ is called ergodic in the:

(a) mean, if $\mathbb{E}[X(t))]$ equals the time average of sample function

$$\mu_X = \mathbb{E}[X(t)] = \lim_{T \to \infty} \frac{1}{T} \int_0^T X(t) \mathrm{d}t, \qquad (1.21)$$

where T is the length of the sample function. Necessary and sufficient conditions for $X(t)$ to be ergodic in the mean are

1. $\mathbb{E}[X(t)] = \text{const}$ and
2. $X(t)$ and $X(t + \tau)$ must become independent as τ reaches ∞.

(b) autocorrelation, if

$$R_X(\tau) = \mathbb{E}[X(t)X(t + \tau)] = \lim_{T \to \infty} \frac{1}{2T} \int_0^T X(t + \tau)X(t) \mathrm{d}t \qquad (1.22)$$

Again, necessary and sufficient conditions for $X(t)$ to be ergodic in the autocorrelation are

1. $\mathbb{E}[X(t + \tau)X(t)]$ is a function of τ only
2. $\lim_{\tau \to \infty} R_X(\tau) = 0$.

Note that an ergodic field is always stationary, but the reverse does not necessarily hold. The concept of ergodicity is of great practical importance, since it allows the estimation of the statistics of a random process from a single time or space record of sufficient length. In most cases though we are only interested a stochastic process to be ergodic in the mean value and in the autocorrelation function, the relevance of stationarity and ergodicity of a stochastic process is depicted in Fig. 1.4.

1.2 Fourier Integrals and Transforms

Any periodic signal can be expressed as a series of harmonically varying quantities called Fourier series. Nonperiodic functions such as stochastic processes can only be expressed as a Fourier series if we consider it to be periodic, with infinite period. This gives rise to the concept of the Fourier integral. The following set of equations give the Fourier integral expression for a nonperiodic function $X(t)$:

$$X(t) = \frac{1}{2\pi} \int_{-\infty}^{\infty} A(\omega)e^{i\omega t} \mathrm{d}\omega \qquad (1.23)$$

$$A(\omega) = \int_{-\infty}^{\infty} X(t)e^{-i\omega t} \mathrm{d}t \qquad (1.24)$$

Fig. 1.4 Forms of
stationarity and their
relevance

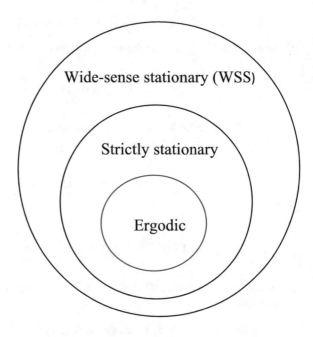

The quantity $A(\omega)$ is called the Fourier transform of $X(t)$. It is in general complex and shows how $X(t)$ is distributed over the frequency range. On its turn, $X(t)$ is said to be the inverse transform of $A(\omega)$. The two quantities $X(t)$ and $A(\omega)$ together form a Fourier transform pair.

1.2.1 Power Spectral Density Function

The power spectrum[3] of a random process $X(t)$ describes how its variance σ_X^2 is distributed over the frequency ω. Consider a stationary ergodic random process $X(t)$ defined over the range $[-\infty, +\infty]$. Such a signal is not periodic and it is thus impossible to define its Fourier transform. It is possible, however, to determine the Fourier transform $A_T(\cdot)$ of a random process $X_T(t)$ which is equal to $X(t)$ over the interval $-\frac{T}{2} \leq t \leq \frac{T}{2}$ and zero at all other times, as follows:

---

[3]The mean-square spectral density function was defined by Norbert Wiener (1894–1964) and independently by Aleksandr Khinchin (1884–1959). However, some of the ideas underlying spectral density preceded these papers. Lord Rayleigh (1842–1919) introduced much earlier the idea that a random process has an autocorrelation and wrote many papers for the spectral representation of random signals.

$$\frac{1}{T}\int_{-\frac{T}{2}}^{\frac{T}{2}} X_T(t)^2 dt = \frac{1}{T}\int_{-\infty}^{+\infty} X_T(t)X_T(t)dt$$

$$= \frac{1}{T}\int_{-\infty}^{+\infty} X_T(t)\left[\frac{1}{2\pi}\int_{-\infty}^{\infty} A_T(\omega)e^{i\omega t}d\omega\right]dt$$

$$= \frac{1}{2\pi T}\int_{-\infty}^{+\infty} A_T(\omega)\left[\int_{-\infty}^{\infty} X_T(t)e^{i\omega t}dt\right]d\omega$$

$$(1.25)$$

If we define the complex conjugate of $A_T(\omega)$ as $A_T^*(\omega) = \int_{-T/2}^{T/2} X_T(t)e^{i\omega t}dt$, then we can write

$$\frac{1}{T}\int_{-\frac{T}{2}}^{\frac{T}{2}} X_T(t)^2 dt = \frac{1}{2\pi T}\int_{-\infty}^{+\infty} A_T(\omega)A_T^*(\omega)d\omega$$

$$= \frac{1}{2\pi T}\int_{-\infty}^{+\infty} |A_T(\omega)|^2 d\omega$$

$$= \frac{1}{\pi T}\int_0^{+\infty} |A_T(\omega)|^2 d\omega$$

$$(1.26)$$

where use is made of the fact that $|A_T(\omega)|^2$ is an even function of ω. If we now let $T \to \infty$, we obtain an expression for the mean-square value (variance) of the original signal $X(t)$:

$$\langle X(t)^2 \rangle = \lim_{T \to \infty} \frac{1}{T}\int_{-\frac{T}{2}}^{\frac{T}{2}} X_T(t)^2 dt$$

$$= \int_{-\infty}^{\infty} \lim_{T \to \infty}\left(\frac{1}{2\pi T}|A_T(\omega)|^2\right)d\omega \qquad (1.27)$$

The spectral density or power density of $X(t)$ is now defined as

$$S(\omega) = \lim_{T \to \infty}\left(\frac{1}{2\pi T}|A_T(\omega)|^2\right) \qquad (1.28)$$

$$= \lim_{T \to \infty} \frac{1}{2\pi T}\left|\int_0^T X(t)e^{-i\omega t}dt\right|^2$$

so that

$$\sigma_X^2 = \int_0^{\infty} S(\omega)d\omega \qquad (1.29)$$

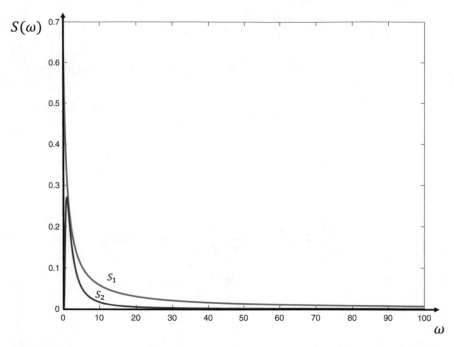

Fig. 1.5 Two power spectrums $S_1(\omega)$ and $S_2(\omega)$ corresponding to Eqs. (1.30) and (1.31), respectively, for $b = \sigma_X = 1$

The power density indicates how the harmonic content of $X(t)$ is spread over the frequency domain. Different realizations of a stationary ergodic random process have the same $S(\omega)$. The power density $S(\omega)$ is thus a constant and deterministic statistical parameter of the random process $X(t)$. Examples of two spectral density functions are depicted in Fig. 1.5, which correspond to the following analytical formulas:

$$S_1(\omega) = \frac{2\sigma_X^2 b^{-1}}{\pi(1 + (\frac{\omega}{b}))} \tag{1.30}$$

$$S_2(\omega) = \frac{2\sigma_X^2 b}{\omega^2} e^{\frac{-2b}{\omega}}, \tag{1.31}$$

where b is a parameter which determines the frequency that $S(\omega)$ has its largest value and its called correlation parameter. The plots correspond to $b = 1$ and $\sigma_X = 1$.

1.2.2 The Fourier Transform of the Autocorrelation Function

There is a direct analytic relation between the autocorrelation function and the spectral density of a stationary random signal. Expressing the autocorrelation as

$$R_X(\tau) = \mathbb{E}[X(t+\tau)X(t)] = \lim_{T\to\infty} \frac{1}{T}\int_{-\frac{T}{2}}^{\frac{T}{2}} X(t+\tau)X(t)\mathrm{d}t \qquad (1.32)$$

its Fourier transform may be written as

$$\int_{-\infty}^{\infty} R_X(\tau)e^{-i\omega\tau}\mathrm{d}\tau = \lim_{T\to\infty}\left[\frac{1}{T}\int_{-\infty}^{\infty}e^{-i\omega\tau}\mathrm{d}\tau\int_{-\frac{T}{2}}^{\frac{T}{2}} X(t+\tau)X(t)\mathrm{d}t\right] \qquad (1.33)$$

Following the procedure described in the previous section, we consider a signal $X_T(t)$ which is equal to $X(t)$ over $-\frac{T}{2} \le t \le \frac{T}{2}$ in order to evaluate the term between square brackets. Omitting the limit $T \to \infty$, the right-hand side of Eq. (1.33) can be written as

$$\frac{1}{T}\int_{-\infty}^{\infty}\mathrm{d}\tau\int_{-\frac{T}{2}}^{\frac{T}{2}} X(t+\tau)X(t)e^{-i\omega\tau}\mathrm{d}t =$$

$$\frac{1}{T}\int_{-\infty}^{\infty}\left[\int_{-\infty}^{\infty} X(t+\tau)X(t)e^{-i\omega\tau}\mathrm{d}t\right]\mathrm{d}\tau =$$

$$\frac{1}{T}\int_{-\infty}^{\infty}\left[\int_{-\infty}^{\infty} X(t+\tau)X(t)e^{-i\omega(t+\tau)}e^{i\omega t}\mathrm{d}t\right]\mathrm{d}\tau = \qquad (1.34)$$

$$\frac{1}{T}\int_{-\infty}^{\infty}\left[\int_{-\infty}^{\infty} X_T(s)X_T(t)e^{-i\omega s}e^{i\omega t}\mathrm{d}t\right]\mathrm{d}s =$$

$$\frac{1}{T}\int_{-\infty}^{\infty} X_T(s)e^{-i\omega s}\mathrm{d}s\int_{-\infty}^{\infty} X_T(t)e^{i\omega t}\mathrm{d}t =$$

$$= \frac{1}{T}A_T^*(\omega)A_T(\omega) = \frac{1}{T}|A_T(\omega)|^2$$

Implementing Eq. (1.34) in (1.33), dividing by 2π and using the definition of the power spectrum density in Eq. (1.28), we find

$$S_X(\omega) = \frac{1}{2\pi}\int_{-\infty}^{\infty} R_X(\tau)e^{-i\omega\tau}\mathrm{d}\tau \qquad (1.35)$$

with the corresponding inverse Fourier Transform given by the following formula:

$$R_X(\tau) = \int_{-\infty}^{\infty} S_X(\omega)e^{i\omega\tau}\mathrm{d}\omega \qquad (1.36)$$

We need to mention at this point that the spectral density function is defined only for stationary processes. The former transform pair of the autocorrelation and power spectrum density function is the statement of the well-known Wiener–Khinchin theorem. Reading that R_X and S_X are real and symmetric functions, Eqs. (1.35) and (1.36) can be written as

$$S_X(\omega) = \frac{1}{\pi} \int_0^\infty R_X(\tau) \cos(\omega\tau) d\tau$$

$$R_X(\tau) = 2 \int_0^\infty S_X(\omega) \cos(\omega\tau) d\omega \tag{1.37}$$

From the above relations, we have $R_X(0) = \int_{-\infty}^\infty S_X(\omega) d\omega = \sigma_X^2$, so that the power spectral density function is finite when $\omega \to \infty$

$$2 \int_{-\infty}^\infty S_X(\omega) d\omega = R_X(0) = \sigma_X^2 < \infty \tag{1.38}$$

1.3 Common Stochastic Processes

1.3.1 Gaussian Processes

As mentioned earlier, a random process can be interpreted as a sequence of random variables. If all these variables are Gaussian, then the random process is said to be Gaussian. A stationary Gaussian process is completely characterized by its mean value μ_X and its autocorrelation function $R_X(\tau)$, since a Gaussian pdf is completely described by its first and second moment. Under linear transformations, Gaussian random processes remain Gaussian.

1.3.2 Markov Processes

In a Markov process, the future behavior, given the past and the present, only depends on the present and not on the past and for this reason is called a memoryless stochastic process. For instance, if $t_1 < t_2, \ldots < t_n$ and X_i are discrete random variables, then the Markov property implies

$$P(X_{t_n} = x_n | X_{t_{n-1}} = x_{n-1}, \ldots, X_{t_1} = x_1) = P(X_{t_n} = x_n | X_{t_{n-1}} = x_{n-1}) \tag{1.39}$$

while in the case X_i are continuous random variables, we have

$$f_{X(t_n)|\{X(t_{n-1})\ldots,X(t_1)\}}(x_n | x_{n-1}, \ldots, x_1) = f_{X(t_n)|X(t_{n-1})}(x_n | x_{n-1}) \tag{1.40}$$

or

$$f_X(x_1, \ldots, x_n; t_1, \ldots, t_n) = f_{X(x_1;t_1)} \prod_{i=2}^n f_{X(t_i)|X(t_{i-1})}(x_i | x_{i-1}) \tag{1.41}$$

Fig. 1.6 State diagram of a
Markov chain

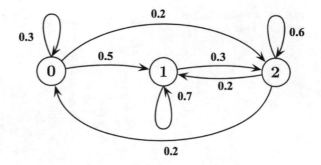

Markov Chain

If time t in the previous equations takes discrete values (finite or countably infinite), then the process is called a Markov chain and is characterized by a set of states S and the transition probabilities p_{ij} between the states, with p_{ij} being the probability that the Markov chain will move from state i to state j. In most cases, we are interested only in systems for which the transition probabilities are independent of time (stationarity assumption)

$$P(X_n = j | X_{n-1} = i) = p_{ij} \tag{1.42}$$

The matrix $\mathbf{P}$ with elements p_{ij} is called the transition probability matrix of the Markov chain

$$\mathbf{P} = \begin{bmatrix} p_{00} & p_{01} & \cdots & p_{0n} \\ p_{10} & p_{11} & \cdots & p_{1n} \\ \vdots & & \vdots & \\ p_{n0} & p_{n1} & \cdots & p_{nn} \end{bmatrix}$$

and has the following important property:

$$\text{for all } i, \quad \sum_{j=1,\dots,n} p_{ij} = 1 \tag{1.43}$$

Sometimes, it is convenient to show $\mathbf{P}$ on a state diagram where each vertex corresponds to a state and each arrow to a nonzero p_{ij}. For example, for the Markov chain with transitional matrix

$$\mathbf{P} = \begin{bmatrix} 0.3 & 0.5 & 0.2 \\ 0 & 0.7 & 0.3 \\ 0.2 & 0.2 & 0.6 \end{bmatrix}$$

we have the diagram of Fig. 1.6.

Fig. 1.7 A typical random
walk on the number line

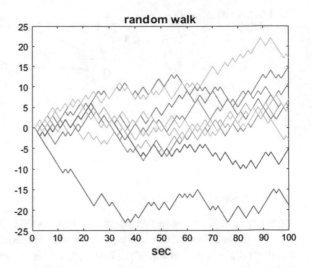

According to Markov's theorem for a finite, aperiodic, irreducible Markov chain
with states $0, 1, \ldots, n$, there exist limiting probabilities π_j, $j = 1, \ldots, n$ such that

$$p_{ij}^{(n)} \to \pi_j, \quad \text{as } n \to \infty \tag{1.44}$$

which are the unique solution of the equations

$$\pi_j = \sum_{i=0}^{n} \pi_i p_{ij} \tag{1.45}$$

under the condition

$$\sum_{i=0}^{n} \pi_i = 1 \tag{1.46}$$

A famous Markov chain is the random walk on the number line (Fig. 1.7). Suppose
that at time $t = 0$, you stand at $x_0 = 0$. You move once every 1 second 1 unit to the
left with probability $p = 0.5$, or 1 unit to the right with probability $p = 0.5$. That is,
$x_1 = 1$ or $x_1 = -1$. Consequently, x_2 will take values in the range of $\{-2, -1, 0, 1, 2\}$
and so forth.

1.3.3   Brownian Process

A famous process that is the continuous analog of the random walk is the Wiener
process or Brownian motion process. Brownian motion process describes the motion

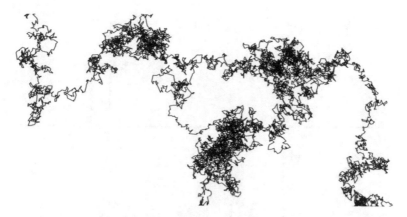

Fig. 1.8 A typical walk in a Brownian motion

of particles suspended in a fluid. Imagine, for instance, a particle immersed in a liquid. Then, due to the random collisions of the particle with the liquid's particles, its trajectory will be similar to a $3D$ random walk (see Fig. 1.8).

Brownian motion is mathematically described with the so-called Wiener process. A Wiener process, $B(t)$, is a continuous time stochastic process, with the following properties:

1. $B(0) = 0$.
2. $B(t)$ is almost surely continuous.[4]
3. $B(t)$ has independent increments.
4. The difference $B(t) - B(s)$ is distributed according to $N(0, t - s)$ pdf for $0 \leq s \leq t$

The derivative of the Brownian motion in the mean-square sense is called white noise.

1.3.4 Stationary White Noise

The stationary white noise is a stochastic process that is characterized by complete randomness. In simple terms, the realizations of the stochastic process at $X(t)$ and $X(t + \tau)$ are totally uncorrelated for any $\tau \neq 0$. The autocorrelation function describing this process is given by

---

[4]A random process $X(t)$ defined on the $\{\Theta, \mathscr{F}, P\}$ probability space is almost surely continuous at t if $P([\omega : \lim_{s \to t} \|X(s, \omega) - X(t, \omega)\| = 0]) = 1$.

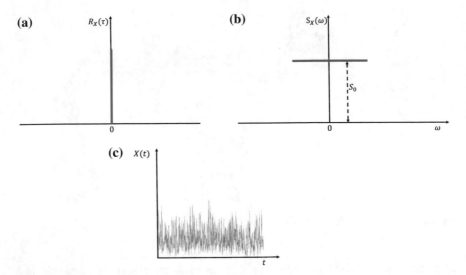

Fig. 1.9 **a** The autocorrelation function, **b** the corresponding power spectrum, and **c** sample realization of a typical stationary white noise

$$R_X(\tau) = 2\pi S_0 \delta_T(\tau), \tag{1.47}$$

where δ_T is the Dirac function defined as

$$\delta_T(\tau) = \begin{cases} 0 & \text{for } \tau \neq 0 \\ \infty & \text{for } \tau = 0 \end{cases}, \quad \int_{-\infty}^{\infty} \delta_T(\tau)\mathrm{d}\tau = 1 \tag{1.48}$$

and S_0 is the corresponding power spectrum which is constant over the frequency domain. A graphical representation of the autocorrelation function, its corresponding power spectrum, and a sample realization of some typical white noise is shown in Fig. 1.9. From the sample, it is obvious that white noise is characterized by complete randomness from point to point.

This idealized white noise is not physically realistic since it corresponds to a process with $\mathbb{E}[X(t)^2] = R_X(0) = \infty$. Instead, if the frequency domain is banded (i.e., $|\omega_1| \leq \omega \leq |\omega_2|$) then we have a band-limited stationary white noise with the following autocorrelation function and power spectrum (Fig. 1.10):

$$R_X(\tau) = 2S_0 \frac{\sin(\omega_2 \tau) - \sin(\omega_1 \tau)}{\tau}$$

$$S(\omega) = \begin{cases} S_0 & \text{for } |\omega_1| \leq \omega \leq |\omega_2| \\ 0 & \text{otherwise} \end{cases}$$

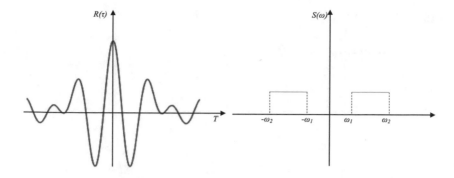

Fig. 1.10 Autocorrelation function and corresponding power spectrum of a stationary band-limited white noise process

1.3.5 Random Variable Case

If a stochastic process can be described by a single random variable, then the autocorrelation function is constant and equal to R_0, while its corresponding power spectrum is a Dirac function at $\omega = 0$ given by

$$S_X(\omega) = R_0 \delta_T(\omega) \tag{1.49}$$

A graphical representation of the autocorrelation function, the power spectrum, and some sample realizations of this process is shown in Fig. 1.11. Sample realizations are straight horizontal lines around the mean value of the process.

1.3.6 Narrow and Wideband Random Processes

Stochastic processes can be characterized by their frequency bandwidth, i.e., the range of frequencies in which the power spectrum has significant mean-square contributions in the process. As shown in Fig. 1.12a, wideband processes have mean-square contributions $S(\omega)$ which are spread out in a wide range of frequencies. Their sample functions appear with incomplete cycles of oscillation and are non-evenly shaped. On the other hand, in narrowband processes $S(\omega)$ is concentrated around some dominant frequency ω_k (Fig. 1.12b) and their sample functions are characterized by uniformity in shape and complete cycles of oscillation.

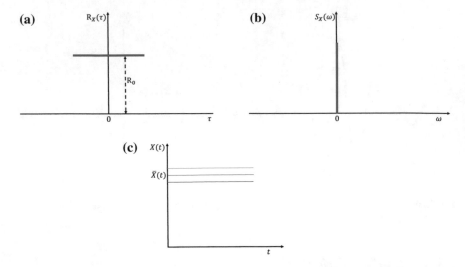

Fig. 1.11 a The autocorrelation function, **b** the corresponding power spectrum, and **c** sample realization of a random variable

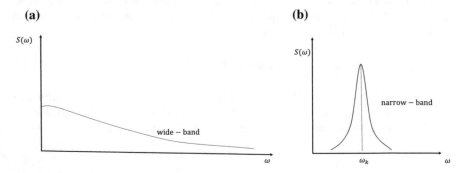

Fig. 1.12 Power spectrum of a **a** wideband stochastic process, **b** narrowband stochastic process

1.3.7   Kanai–Tajimi Power Spectrum

The stationary Kanai and Tajimi model is well-known and favorite process for many researchers and engineers, which is widely used in the field of earthquake engineering for the description of the seismic ground motion acceleration. Its power spectrum is given by

$$S(\omega) = \frac{1 + 4\zeta_g^2(\omega/\omega_g)^2}{[1 + (\omega/\omega_g)^2]^2 + 4\zeta_g^2(\omega/\omega_g)^2} S_0, \qquad (1.50)$$

where S_0 is a constant related to the intention of the accelerogram, thus depending on the earthquake magnitude and distance from the epicenter and ω_g, ζ_g the characteristic ground frequency and damping, respectively (Fig. 1.13). The corresponding

Fig. 1.13 The Kanai–Tajimi
power spectrum

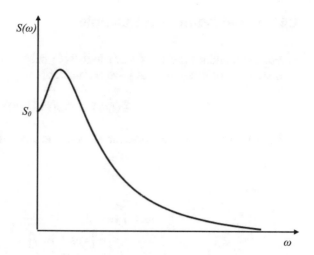

autocorrelation function is defined as

$$R(\tau) = \frac{S_0\omega_g^2}{4\omega_g\xi_g}\sqrt{\frac{1+8\xi_g^2}{1-\xi_g^2}}e^{-\omega_g\xi_g|\tau|}\cos(\omega_g^d|\tau|-\phi), \qquad (1.51)$$

where

$$\omega_g^d = \omega_g\sqrt{1-\xi_g^2}, \quad \phi = \tan^{-1}\frac{\xi_g(1-4\xi_g^2)}{\sqrt{1-\xi_g^2(1+4\xi_g^2)}} \qquad (1.52)$$

Kanai–Tajimi spectrum exhibits a nor-ergodic constant term for the autocorrelation since it has power S_0 at $\omega = 0$. This creates singularities of the ground velocities and displacements time histories which become unbounded in variance. For this reason, Clough and Penzien suggested to multiply the Kanai–Tajimi spectrum by the following correction term:

$$S^*(\omega) = \frac{(\omega/\omega_c)^4}{[1+(\omega/\omega_c)^2]^2 + 4\zeta_c^2(\omega/\omega_c)^2}, \qquad (1.53)$$

which is a second-order high pass filter which helps to overcome the singularities of Kanai–Tajimi power spectrum at small frequencies. ω_c and ζ_c in Eq. (1.53) are filtering parameters.

1.4   Solved Numerical Examples

1. For the random process $X(t, \theta)$ with $\xi(\theta)$ being a random variable uniformly distributed over the interval $[-\pi, \pi]$ and let

$$X(t, \theta) = \cos(t + \xi(\theta)) \qquad (1.54)$$

Find the mean, variance, autocorrelation, and autocovariance of $X(t, \theta)$.

Solution:

- **mean**

$$
\begin{aligned}
m_x(t) &= \\
&= \mathbb{E}\big[\cos(t + \xi)\big] \\
&= \frac{1}{2\pi} \int_{-\pi}^{\pi} \cos(t + \xi)\,\mathrm{d}\xi = 0 \qquad (1.55)
\end{aligned}
$$

- **autocovariance**

$$
\begin{aligned}
C_X(t_1, t_2) &= \\
&= \mathbb{E}\big[\cos(t_1 + \xi)\cos(t_2 + \xi)\big] \\
&= \frac{1}{2\pi} \int_{-\pi}^{\pi} \frac{1}{2}\big\{\cos(t_1 - t_2) + \cos((t_1 + t_2) + \xi)\big\}\mathrm{d}\xi \\
&= \frac{1}{2}\cos(t_1 - t_2) \qquad (1.56)
\end{aligned}
$$

- **autocorrelation**

$$
\begin{aligned}
C_X(t_1, t_2) &= R_X(t_1, t_2) - \mu_X(t_1)\mu_X(t_2) \Leftrightarrow \\
R_X(t_1, t_2) &= C_X(t_1, t_2) \\
&= \frac{1}{2}\cos(t_1 - t_2) \qquad (1.57)
\end{aligned}
$$

- **variance**

$$
\begin{aligned}
\mathrm{Var}[X(t)] &= C_X(t, t) \\
&= \frac{1}{2}\cos(t - t) \\
&= \frac{1}{2} \qquad (1.58)
\end{aligned}
$$

2. Consider the stationary random process given by

$$X(t) = A\sin(\omega t + \varphi) \qquad (1.59)$$

where A and ω are constants and φ are random phase angles uniformly distributed between 0 and 2π

$$f_\Phi(\varphi) = \begin{cases} \frac{1}{2\pi} & \text{for} 0 \leq \varphi \leq 2\pi \\ 0 & \text{otherwise} \end{cases}$$

Calculate the autocorrelation function of $X(t)$ and show that it is a function of $\tau = t_1 - t_2$.

Solution:

$$\mathbb{E}[X(t)] = \int_{-\infty}^{+\infty} A\sin(\omega t + \varphi) f_\Phi(\varphi) d\varphi$$

$$= \int_0^{2\pi} A\sin(\omega t + \varphi) \frac{1}{2\pi} d\varphi = -\frac{A}{2\pi}[\cos(\omega t + \varphi)]_0^{2\pi} = 0$$

$$\begin{aligned} R_X(t_1, t_2) &= \mathbb{E}[X(t_1)X(t_2)] \\ &= \mathbb{E}[A\sin(\omega t_1 + \varphi)A\sin(\omega t_2 + \varphi)] \\ &= \int_0^{2\pi} A^2 \sin(\omega t_2 + \varphi)\sin(\omega t_1 + \varphi) \frac{1}{2\pi} d\varphi \\ &= \frac{A^2}{2\pi} \int_0^{2\pi} \frac{1}{2}\left[\cos(\omega t_1 + \varphi - \omega t_2 - \varphi) - \cos(\omega t_1 + \varphi + \omega t_2 + \varphi)\right] d\varphi \\ &= \frac{A^2}{4\pi} \int_0^{2\pi} \cos[(\omega(t_1 - t_2)]d\varphi - \frac{A^2}{4\pi} \int_0^{2\pi} \cos[(\omega(t_1 + t_2) + 2\varphi]d\varphi \\ &= \frac{A^2}{4\pi}[\cos(\omega(t_1 - t_2))]2\pi - 0 \\ &= \frac{A^2}{2}[\cos(\omega(t_1 - t_2))] = \frac{A^2}{2}[\cos(\omega\tau)] = R_X(\tau) \end{aligned}$$

3. A random process $X(t, \theta)$ is given by

$$X(t) = A(\theta) \cdot \cos(\omega t) + B(\theta)\sin(\omega t),$$

where ω is a constant and $A(\theta)$ and $B(\theta)$ are independent, identically distributed random variables with zero means and standard deviations σ. Show that $X(t, \theta)$ is a stationary random process.

Solution:

$$\begin{aligned} \mathbb{E}[X(t, \theta)] &= \mathbb{E}[A\cos(\omega t) + B\sin(\omega t)] \\ &= \mathbb{E}[A]\cos(\omega t) + \mathbb{E}[B]\sin(\omega t) = 0 \end{aligned}$$

$$R_X(t_1, t_2) = \mathbb{E}[X(t_1)X(t_2)] = \mathbb{E}\Big[\big(A\cos(\omega t_1) + B\sin(\omega t_1)\big)...$$

$$\big(A\cos(\omega t_2) + B\sin(\omega t_2)\big)\Big]$$

$$= \mathbb{E}\Big[\big(A^2\cos(\omega t_1)\cos(\omega t_2) + B^2\sin(\omega t_1)\sin(\omega t_2) + ...$$

$$+ AB\big(\sin(\omega t_1)\cos(\omega t_2) + \cos(\omega t_1)\sin(\omega t_2)\big)\Big]$$

$$= \mathbb{E}\Big[A^2\Big]\cos(\omega t_1)\cos(\omega t_2) + \mathbb{E}\Big[B^2\Big]\sin(\omega t_1)\sin(\omega t_2) + ...$$

$$+ \mathbb{E}\Big[AB\Big]\big(\sin(\omega t_1)\cos(\omega t_2) + \cos(\omega t_1)\sin(\omega t_2)\big)$$

Using $\mathbb{E}[A^2] = \mathbb{E}[B^2] = \sigma^2$ and $\mathbb{E}[AB] = 0$, since A and B are independent random variables, the autocorrelation function becomes

$$R_X(t_1, t_2) = \sigma^2\cos(\omega(t_2 - t_1)) = \sigma^2\cos(\omega\tau) = R_X(\tau)$$

Therefore, $X(t, \theta)$ is a stationary random process.

4. Show that the following autocorrelation function is not realistic:

$$R_X(\tau) = \begin{cases} R_0 & \text{for}|\tau| \leq \tau_c \\ 0 & \text{otherwise} \end{cases}$$

Solution:

The corresponding power spectral density function is given by

$$S_X(\omega) = \frac{1}{2\pi}\int_{-\infty}^{+\infty} R_X(\tau)e^{-i\omega t}d\tau$$

$$= \frac{1}{2\pi}\int_{-\infty}^{+\infty} R_0\cos(\omega\tau)d\tau$$

$$= \frac{R_0}{\pi\omega}\sin(\omega\tau_c)$$

The above power spectral density function can take negative values which violates the nonnegative property. Therefore, this autocorrelation function is not realistic.

5. Show that the stationary random process $X(t, \theta) = A\sin(\omega t + \varphi(\theta))$, where $A, \omega = $ constants and Φ is a random phase angle, uniformly distributed between 0 and 2π is ergodic in the first moment and in correlation.

Solution:

It has been shown in **Example 2** that $\mathbb{E}[X(t)] = 0$ and $R_X(\tau) = \frac{A^2}{2}\cos(\omega\tau)$. In order for the random process to be ergodic in the mean and in the autocorrelation, we need to show that the following relations stand:

$$\mathbb{E}[X(t,\theta)] = \lim_{T\to\infty} \frac{1}{T} \int_0^T X(t,\theta)dt$$

$$\mathbb{E}[X(t,\theta)X(t+\tau,\theta)] = \lim_{T\to\infty} \frac{1}{T-\tau} \int_0^{T-\tau} X(t,\theta)X(t+\tau,\theta)dt$$

If the first one is satisfied, then the random process is ergodic in the mean while, if the second one is satisfied, then the process is ergodic in the autocorrelation. Thus, the integral of the first equation gives

$$\lim_{T\to\infty} \frac{1}{T} \int_0^T X(t)dt = \lim_{T\to\infty} \frac{1}{T} \int_0^T A\sin(\omega t + \varphi)dt$$

$$= \lim_{T\to\infty} \frac{1}{T}\frac{A}{\omega}\left[-\cos(\omega t + \varphi)\right]_0^T$$

$$= \lim_{T\to\infty} \frac{A}{T\omega}\left[-\cos(\omega T + \varphi) + \cos(\varphi)\right]$$

$$= 0 = \mathbb{E}[X(t,\theta)]$$

while the integral of the second equation is

$$\lim_{T\to\infty} \frac{1}{T-\tau} \int_0^{T-\tau} X(t)X(t+\tau)dt =$$

$$= \lim_{T\to\infty} \frac{1}{T-\tau} \int_0^{T-\tau} A\sin(\omega t + \varphi)A\sin(\omega(t+\tau)+\varphi)dt$$

$$= \lim_{T\to\infty} \frac{1}{T-\tau}A^2 \int_0^{T-\tau} \frac{1}{2}[\cos(\omega\tau) - \cos(2\omega t + \omega\tau + 2\varphi)]dt$$

$$= \lim_{T\to\infty} \frac{A^2}{2}\frac{1}{T-\tau}\left\{(T-\tau)\cos(\omega\tau) - \frac{1}{2\omega}\left[\sin(2\omega t + \omega\tau + 2\varphi)\right]_0^{T-\tau}\right\}$$

$$= \lim_{T\to\infty} \frac{A^2}{2}\cos(\omega\tau) - \lim_{T\to\infty} \frac{A^2}{4}\frac{1}{T-\tau}\left[\sin(2\omega T - \omega\tau + 2\varphi)...\right.$$

$$\left.... - \sin(\omega\tau + 2\varphi)\right] = \frac{A^2}{2}\cos(\omega\tau)$$

6. If $X(t,\theta)$ is a WSS random process and A is the area under $X(t,\theta)$ in the time frame $[t_1, t_2]$, calculate the mean and the variance of the random variable A.

Solution:

The random variable A can be defined as

$$A = \int_{t_1}^{t_2} X(t)dt$$

then, by definition the mean value can be calculated as

$$\mathbb{E}[A] = \int_{t_1}^{t_2} \mathbb{E}[X(t)]dt = c(t_2 - t_1) = \text{const.}$$

Since $\mathbb{E}[X(t)] = \text{const.}$ the variance can be estimated as

$$A^2 = \int_{t_1}^{t_2} \int_{t_1}^{t_2} X(t_1)X(t_2)dt_1 dt_2 \Rightarrow \qquad (1.60)$$

$$\mathbb{E}[A^2] = \int_{t_1}^{t_2} \int_{t_1}^{t_2} \mathbb{E}[X(t_1)X(t_2)]dt_1 dt_2$$

$$= \int_{t_1}^{t_2} \int_{t_1}^{t_2} R(t_1, t_2)dt_1 dt_2$$

7. The simply supported beam of Fig. 1.14 is subjected to a stochastically distributed load $q(x)$, modeled through a stochastic field with mean q_0 and autocorrelation function $R_q(x_1, x_2)$. Calculate the mean and the variance of the reaction at support A.

Solution:

The reaction at support A can be estimated by

$$V_A = \frac{1}{L} \int_0^L q(x)(L - x)dx \Rightarrow$$

$$\Rightarrow \mathbb{E}[V_A] = \frac{1}{L} \int_0^L \mathbb{E}[q(x)](L - x)dx = q_0 L/2$$

The variance of the reaction V_A is

$$V_A^2 = \frac{1}{L^2} \int_0^L \int_0^L q(x_1)q(x_2)(L - x_1)(L - x_2)dx_1 dx_2 \Rightarrow$$

$$\Rightarrow \mathbb{E}[V_A^2] = \frac{1}{L^2} \int_0^L \int_0^L \mathbb{E}[q(x_1)q(x_2)(L - x_1)(L - x_2)]dx_1 dx_2$$

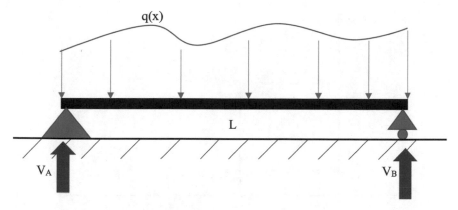

Fig. 1.14 A simply supported beam under a stochastically distributed load

$$= \int_0^L \int_0^L g(x_1, x_2) R_q(x_1, x_2) dx_1 dx_2$$

where $g(x_1, x_2) = \dfrac{(L - x_1)(L - x_2)}{L^2}$ is a deterministic function.

1.5 Exercises

1. Let Z be a random variable and define the stochastic process $X(t) = Z$. Show that $X(t)$ is a strictly stationary process.
2. Let $Y_0, Y_1, \ldots, Y_m$ and $X_0, X_1, \ldots, X_m$ be uncorrelated random variables with mean zero and variances $\mathbb{E}[Y_i^2] = \sigma_i^2$ and $\mathbb{E}[X_i^2] = \sigma_i^2$ with $i = 1, \ldots, m$. For the stochastic process

$$Z_n = \sum_{k=0}^m (Y_k \cos(n\omega_k) + X_k \sin(n\omega_k)),$$

where ω_k are frequencies uniformly distributed in $[0, \pi]$ and $n = 0, \pm 1, \pm 2, \ldots$, calculate the mean and covariance and show that it is weakly stationary.
3. For $Z(t) = aX(t) + bY(t) + c$ where and a, b, c are constants and $X(t), Y(t)$ are stationary stochastic processes with means μ_X, μ_Y, respectively, autocorrelation functions $R_X(\tau), R_Y(\tau)$, and cross-correlation function $R_{XY}(\tau)$, evaluate the autocorrelation and power spectrum. Use the fact that the mean value operator is linear and the symmetry property of the cross-correlation function.
4. Find the autocorrelation function $R_X(\tau)$ of a stochastic process $X(t)$ if (a) $S(\omega) = 1/(4 + \omega^5)$ and (b) $S(\omega) = 2/(3 + \omega^2)^3$.

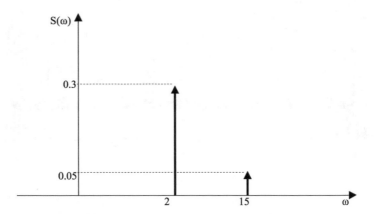

Fig. 1.15 Power spectrum of a stochastic process

5. Consider the following stochastic processes $X(t) = 35(\sin 5t + \phi)$where ϕ is a random variable uniformly distributed in the range $[0, 2\pi]$. Estimate both numerically and analytically their autocorrelation function and the corresponding power spectral density function.

6. If $X(t)$ is a WSS process with $\mathbb{E}[X(t)] = 50$ and $R(\tau) = e^{-|\tau|/465}$ calculate the variance and covariance of $X(100)$ and $X(262)$.

7. If the power spectrum of a stochastic process $X(t)$ has the shape shown in Fig. 1.15, provide possible sample realizations of the process.

8. For the autocorrelation function of exponential type $R_X(\tau) = e^{-\frac{|\tau|}{b}}$ with b being the correlation length parameter, compute and then plot the corresponding power spectrum using inverse fourier transform for $b = 1, 50$, and 500. Give an estimate of the shape of the sample functions of a process that has the corresponding correlations structures. The variance of the process is equal to one.

9. For the **Example 7**, give the expressions of the bending moment $M(x)$ as well as the shear force $Q(x)$ and calculate their mean value and variance. Are these fields stationary?

10. Generate a large number of samples of a random walk $X(t_n)$ that has the following properties:

   a. Each sample starts from zero at $t_0 = 0$;
   b. Consists of 1000 time steps
   c. The probability at each step to go up or down is equal.

   Prove the following: (i) $\mathbb{E}[X(t_n)] = 0$ and (ii) $\text{Var}[X(t_n)] = n$.

Chapter 2
Representation of a Stochastic Process

Numerical analysis of engineering problems in a probabilistic framework requires a representation of some stochastic processes, used for the description of the uncertainties involved in the problem. To this purpose, a continuous mD-nV, in the general case, stochastic process $X(t, \theta)$ needs to be represented by discrete values X_i at some discretization points $i = 1, \ldots, N$. So, the main question becomes: "how can one determine the optimal approximation process (or field) $\hat{X}(\cdot)$ which will best describe the original process $X(\cdot)$ with the minimum number of random variables $\{X_i\}$?", i.e.,

$$X(t, \theta) \approx \hat{X}(t, \theta) = \{X_i\} \tag{2.1}$$

Since, as will be shown in Chaps. 3 and 4, the computational effort in stochastic and/or reliability analysis problems is proportional to the number of random variables, the answer to this question is crucial and comes with respect to some error estimator as will be discussed later. Along these lines, all SFEM approaches are based on some kind of representation of the stochastic processes as a series of random variables. Without loss of generality this description is given for a 1D-1V stochastic process. The discretization methods can be generally categorized to the following groups:

1. **Point discretization** methods, where the random variables X_i are values of $X(t, \theta)$ at some given points t_i.

$$\hat{X}(t, \theta) = \{X_i\} = \{X(t_i, \theta)\} \tag{2.2}$$

2. **Average discretization** methods, where the random variables X_i are weighted integrals of $X(t, \theta)$ over a domain Ω_i.

$$\hat{X}(t, \theta) = \{X_i\} = \left\{ \int_{\Omega_i} X(t, \theta) c(t) \mathrm{d}\Omega_i, \quad t \in \Omega_i \right\} \tag{2.3}$$

© Springer International Publishing AG 2018
V. Papadopoulos and D.G. Giovanis, *Stochastic Finite Element Methods*,
Mathematical Engineering, https://doi.org/10.1007/978-3-319-64528-5_2

where $c(\cdot)$ are corresponding weights.

3. **Interpolation** methods, where the stochastic process can be approximated by interpolation at some points. Usually it is implemented in combination with some point discretization method.

4. **Series expansion** methods, where the stochastic process is represented as a truncated finite series, expressed as the decoupled product of some random variables with deterministic spatial functions. The most widely used methods of this category are the Karhunen–Loève expansion (KL) and the spectral representation (SR) methods. Both these methods belong to the family of spectral methods and $\hat{X}(t)$ is expressed as an infinite series as

$$\hat{X}(t, \theta) = \sum_{j=1}^{\infty} g_j(t) \xi_j(\theta) \tag{2.4}$$

where $\{\xi_j(\theta)\}$ are random variables considered as the coefficients of the series and $g_j(t)$ are deterministic functions over which the original field is projected.

2.1   Point Discretization Methods

The point discretization methods represent a stochastic process $\hat{X}(t, \theta)$ as discrete random variables at one or more points t_i. The value of the process at point i is then given by

$$\hat{X}(t_i, \theta) = X_i = X(t_i) \tag{2.5}$$

where t_i are the coordinates of point i. The mean value and the variance of the random variable X_i is the sample of the stochastic process at that point. The correlation between two points can be approximated from the autocorrelation function

$$R_{t_i, t_j} = R_X(t_i, t_j) \tag{2.6}$$

All the point discretization methods described next have three advantages

- The covariance matrix can be easily calculated.
- The covariance matrix is positive definite.
- The discrete random variables and the stochastic process have the same probability distribution function and for this reason the simulation process is independent of the type (pdf) of the process.

The main disadvantage of point discretization methods in the framework of stochastic finite element applications is that in order for the properties of the process to remain constant within a finite element, the size of all the elements must be small enough (fraction of the correlation length). As explained later (example 1 in Sect. 3.6), this

condition leads in certain cases to a relative dense finite element mesh which increases
the computational cost.

2.1.1 Midpoint Method

The midpoint method (MP) is implemented in a discretized domain and uses the
value of the stochastic process at the centroid $t_i^\star$ of the element i

$$\hat{X}(t_i, \theta) = X_i = X(t_i^\star) \tag{2.7}$$

where the $t_i^\star$ coordinates are obtained from the spatial coordinates of the nodes $t_j^{(i)}$
of the element as:

$$t_i^\star = \frac{1}{N} \sum_{k=1}^{N} t_j^{(i)} \tag{2.8}$$

where N is the number of nodes of element i. For a finite element mesh consisting of
N_e elements the stochastic process $\hat{X}(t, \theta)$ is defined by the random vector $\{X_i\} = [X(t_1^\star), X(t_2^\star), \ldots X(t_{N_e}^\star)]$.

2.1.2 Integration Point Method

The integration method can be considered as an extension of the MP method by
associating a single random variable to each Gauss point of the finite element model
instead of the centroid of each element. The main drawback of this method is that
the total number of random variables involved increases dramatically with the size
of the problem at hand.

2.1.3 Average Discretization Method

The average discretization method defines the approximated process $\hat{X}(t, \theta)$ in each
finite element i as the average of the original process over the volume Ω_i of the
element

$$\hat{X}(t, \theta) = X_i = \frac{1}{\Omega_i} \int_{\Omega_i} X(t, \theta) d\Omega_i, \quad t \in \Omega_i, \tag{2.9}$$

In a finite element mesh consisting of N_e elements, the approximated process $\hat{X}(t, \theta)$
is defined by the collection $\{X_i\} = [X_1, \ldots, X_{N_e}]$ of these N_e random variables. The

mean and covariance matrix of $\{X_i\}$ are computed from the mean and covariance function of $X(t, \theta)$ as integrals over the domain Ω_i.

2.1.4 Interpolation Method

The interpolation method approximates the stochastic process $\hat{X}(t, \theta)$ in an element Ω_i using the values of the process at q nodal coordinates t and corresponding shape functions $N(t)$ as follows:

$$\hat{X}(t, \theta) = X_i = \sum_{j=1}^{q} N_j(t) X(t_j) \quad t \in \Omega_i \qquad (2.10)$$

The nodal points do not necessarily coincide with the nodes of the element and the shape functions N_j can be chosen independently of the shape functions of the finite element model. The mean value and variance of the approximated field within each element are given by

$$\mathbb{E}\big[\hat{X}(t, \theta)\big] = \sum_{j=1}^{q} N_j(t) \mathbb{E}[X(t_j)], \quad t \in \Omega_i \qquad (2.11)$$

$$\mathrm{Var}\big[\hat{X}(t, \theta)\big] = \sum_{j=1}^{q} \sum_{k=1}^{q} N_j(t) N_k(t) R_X(t_j, t_k), \quad t \in \Omega_i \qquad (2.12)$$

Each realization of $\hat{X}(\cdot)$ is a continuous function over Ω_i which is an advantage over the midpoint method. The main disadvantage of this method is that due to the interpolation, the marginal pdf of $\hat{X}(\cdot)$ is not fully consistent to the one of $X(\cdot)$.

2.2 Series Expansion Methods

2.2.1 The Karhunen–Loève Expansion

The Karhunen–Loève expansion[1] of a zero-mean random process $X(t, \theta)$ is based on the spectral decomposition of its covariance function defined as

$$C_X(t_i, t_j) = \sigma_X(t_i) \sigma_X(t_j) \rho(t_i, t_j) \qquad (2.13)$$

---

[1]Named after the probabilist and mathematical statisticians Kari Karhunen (1907–1979) and Michel Loève (1915–1992).

where ρ is the correlation coefficient. By definition, $C_X(t_i, t_j)$ is bounded, symmetric and has the following spectral or eigen-decomposition:

$$C_X(t_i, t_j) = \sum_{n=1}^{\infty} \lambda_n \varphi_n(t_i) \varphi_n(t_j) \tag{2.14}$$

where φ_n and λ_n are orthogonal deterministic eigenfunctions and eigenvalues of the covariance function, respectively, derived from the solution of the homogeneous Fredholm integral equation of the second kind for the covariance kernel

$$\int_{\mathscr{D}} C_X(t_i, t_j) \varphi_n(t_j) dt_j = \lambda_n \varphi_n(t_i) \tag{2.15}$$

where $\mathscr{D}$ is the domain in which the stochastic processes is defined. The key to KL expansion is to obtain the eigenvalues and eigenfunctions by solving Eq. (2.15). Because an analytical solution of Eq. (2.15) is tractable only is special cases, in general a numerical solution is the only resort.

The eigenfunctions form a complete orthogonal set satisfying the equation

$$\int_{\mathscr{D}} \varphi_k(t) \varphi_l(t) dt = \delta_{kl} \tag{2.16}$$

where δ_{kl} is the Kronecker-delta function. Any realization of $X(t, \omega)$ can thus be expanded over this basis as follows:

$$X(t, \theta) = \sum_{i=1}^{\infty} \sqrt{\lambda_n} \varphi_n(t) \xi_n(\theta), \quad t \in \mathscr{D} \tag{2.17}$$

where $\xi_n(\theta)$ is a set of uncorrelated random variables with mean $\mathbb{E}\big[\xi_n(\theta)\big] = 0$ and covariance function $\mathbb{E}\big[\xi_k(\theta)\xi_l(\theta)\big] = \delta_{kl}$ which can be expressed as

$$\xi_n(\theta) = \frac{1}{\sqrt{\lambda_n}} \int_{\mathscr{D}} X(t, \theta) \varphi_n(t) dt \tag{2.18}$$

Equation (2.17) is known to converge in the mean square sense for any distribution of $X(t, \theta)$. The KL expansion of a Gaussian process has the property that $\xi_n(\theta)$ are independent standard normal variables. For practical implementation, the series is approximated by a finite number of terms M, giving

$$X(t, \theta) \approx \hat{X}(t, \theta) = \sum_{n=1}^{M} \sqrt{\lambda_n} \varphi_n(t) \xi_n(\theta) \tag{2.19}$$

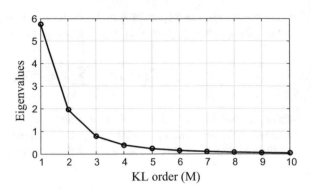

Fig. 2.1 Decaying eigenvalues from the solution of the Fredholm integral of the second kind for $M = 10$

The corresponding covariance function is then approximated by

$$\hat{C}_X(t_i, t_j) = \sum_{n=1}^{M} \lambda_n \varphi_n(t_i) \varphi_n(t_j) \tag{2.20}$$

Ghanem and Spanos (1991) demonstrated that this truncated series is optimal in the mean square since the eigenvalues λ_n of Eq. (2.19) are converging fast to zero (Fig. 2.1). Thus, the choice of the covariance eigenfunction basis $\{\varphi_n(t)\}$ is optimal in the sense that the mean square error resulting from a truncation after the M-th term is minimized.

The variance error e_{var} after truncating the expansion in M terms can be easily computed as

$$e_{var} = \text{Var}\big[X(t, \theta) - \hat{X}(t, \theta)\big] = \sigma_X^2 - \sum_{n=1}^{M} \lambda_n \varphi_n^2(t) \tag{2.21}$$

The righthand side of the above equation means that the KL expansion always under-represents the true variance of the field.

Analytical Solution of the Integral Eigenvalue Problem

For some types of covariance functions, the Fredholm integral equation of Eq. (2.15) can be differentiated twice with respect to t_j. The resulting differential equation then can be solved analytically in order to yield the eigenvalues. An example of this class is the first-order Markov process (see Sect. 1.3.2) defined in the symmetrical domain $\mathscr{D} = [-a, a]$ and has the following covariance function:

$$C(t_i, t_j) = \sigma^2 \exp\left(\frac{|t_i - t_j|}{b}\right) \tag{2.22}$$

where σ^2 is the variance and b is a correlation length parameter. For $\sigma = 1$ Eq. (2.15) can be written as

$$\int_{-a}^{a} \exp\left(\frac{|t_i - t_j|}{b}\right)\varphi_n(t_j)dt_j = \lambda_n \varphi(t_i) \tag{2.23}$$

The eigenvalues and the eigenfunctions in Eq. (2.23) can easily be estimated as follows:

- For $n = $ odd,

$$\lambda_n = \frac{2b}{1 + \omega_n^2 b^2}, \quad \varphi_n(t) = c_n \cos(\omega_n t) \tag{2.24}$$

where c_n is given by

$$c_n = \frac{1}{\sqrt{a + \frac{\sin(2\omega_n a)}{2\omega_n}}}$$

and ω_n is obtained from the solution of

$$\frac{1}{b} - \omega_n \tan(\omega_n a) = 0 \quad \text{in the range} \left[(n-1)\frac{\pi}{a}, \left(n - \frac{1}{2}\right)\frac{\pi}{a},\right] \tag{2.25}$$

- For $n \geq 2$ and $n = $ even,

$$\lambda_n = \frac{2b}{1 + \omega_n^2 b^2}, \quad \varphi_n(t) = l_n \sin(\omega_n t) \tag{2.26}$$

with

$$l_n = \frac{1}{\sqrt{a - \frac{\sin(2\omega_n a)}{2\omega_n}}} \tag{2.27}$$

and ω_n being the solution of

$$\frac{1}{b}\tan(\omega_n a) + \omega_n = 0 \quad \text{in the range} \left[\left(n - \frac{1}{2}\right)\frac{\pi}{a}, n\frac{\pi}{a},\right] \tag{2.28}$$

The solution of the Fredholm integral equation of the second kind is analytically given in the section of solved numerical examples. In this part, we need to mention that the aforementioned solutions stand for the cases of symmetrical domains $\mathcal{D}$. If $\mathcal{D}$ is not symmetrical, e.g., $\mathcal{D} = [t_{min}, t_{max}]$, then a shift parameter $T = (t_{min} + t_{max})/2$ is required in order to obtain the solution and the Fredholm integral equation is solved over the domain

$$\mathcal{D}' = \mathcal{D} - T = \left[\frac{t_{min} - t_{max}}{2}, \frac{t_{max} - t_{min}}{2}\right] \tag{2.29}$$

Thus, we have

$$\hat{X}(t, \theta) = \sum_{n=1}^{M} \sqrt{\lambda_n} \varphi_n(t - T) \xi_n(\theta) \tag{2.30}$$

Inspection of Eq. (2.23) indicates that the quality of the simulated stochastic field is affected by the length of process relatively to the correlation parameter b and the number of KL terms M. A detailed investigation of these sensitivities was preformed in Huang et al. (2001) which revealed the following important properties:

1. A low value of a/b implies a highly correlated process and hence, a relative small number of random variables are required to represent the stochastic process. Correspondingly, fewer number of terms in the KL expansion are needed for a qualitative representation.
2. The faster the autocorrelation function converges to zero, the wider is the corresponding power spectral density hence, a greater number of terms is required to sufficiently represent the underlying process by KL.
3. For a given M, the accuracy decreases as the fraction a/b increases.

For a fixed M, analytical KL gives significantly better results than numerical KL. A short description of the numerical KL is following.

Numerical Solution of the Integral Eigenvalue Problem

For random processes where the analytical solution of the Fredholm integral equation is intractable, a numerical solution is necessary. One major category of such solution schemes are the expansion methods such as the Galerkin, the collocation and the Rayleigh–Ritz methods. Galerkin methods are essentially error minimization schemes with respect to some residual calculated over the entire domain of the solution. Assuming that each eigenfunction $\varphi_n(t)$ of $C_X(t_i, t_j)$ may be represented by its expansion over a polynomial basis $\{h_i(\cdot)\}$, defined in the solution space, as

$$\varphi_n(t) = \sum_{i=1}^{\infty} d_i^n h_i(t) \tag{2.31}$$

where d_i^n are unknown coefficients to be estimated, the Galerkin procedure targets to an optimal approximation of the eigenfunctions $\varphi_n(\cdot)$ after truncating the above series in N terms and computing the residual as

$$\varepsilon_N(t) = \sum_{i=1}^{N} d_i^n \left[\int_{\mathscr{D}} C_X(t_i, t_j) h_i(t_j) \mathrm{d}t_j - \lambda_j h_i(t) \right] \tag{2.32}$$

Requiring the residual to be orthogonal to the space spanned by the same basis we get

$$< \varepsilon_N, h_j > := \int_{\mathscr{D}} \varepsilon_N(t) h_j(t) \mathrm{d}t = 0, \quad j = 1, \ldots, N \tag{2.33}$$

which leads to the following matrix eigenvalue equation:

$$\mathbf{CD} = \mathbf{\Lambda BD} \qquad (2.34)$$

where

$$\mathbf{B}_{ij} = \int_{\mathcal{D}} h_i(t)h_j(t)\mathrm{d}t \qquad (2.35)$$

$$\mathbf{C}_{ij} = \int_{\mathcal{D}}\int_{\mathcal{D}} C_X(t_i, t_j)h_i(t_j)\mathrm{d}t_j \qquad (2.36)$$

$$\mathbf{D}_{ij} = d_i^j \qquad (2.37)$$

$$\mathbf{\Lambda}_{ij} = \delta_{ij}\lambda_j \qquad (2.38)$$

where $\mathbf{C}$, $\mathbf{D}$, $\mathbf{B}$ and $\mathbf{\Lambda}$ are $N \times N$-dimensional matrices. This generalized algebraic eigenvalue problem of Eq. (2.34) can be solved for $\mathbf{D}$ and $\mathbf{\Lambda}$ and with backsubstitution we can estimate the eigenfunctions of the covariance kernel. This solution scheme can be implemented using piecewise polynomials for the basis $\{h_i(\cdot)\}$ of the expansion.

2.2.2 Spectral Representation Method

The spectral representation method was proposed by Shinozuka and Deodatis (1991) and generates sample functions that are ergodic in the mean value and autocorrelation. Its main property is that it expands the stochastic field to a series of trigonometric functions with random phase angles. For a zero-mean, one-dimensional stationary stochastic process $X(t, \theta)$ with autocorrelation function $R_X(t_i, t_j)$ and two-sided power spectral function $S_X(\omega)$ we can define two mutually orthogonal real-valued processes $u(\omega)$ and $v(\omega)$ with corresponding orthogonal steps $\mathrm{d}u(\omega)$ and $\mathrm{d}v(\omega)$, respectively, such that

$$X(t, \theta) = \int_0^\infty \left[\cos(\omega t)\mathrm{d}u(\omega) + \sin(\omega t)\mathrm{d}v(\omega) \right] \qquad (2.39)$$

The processes $u(\omega)$ and $v(\omega)$ and their corresponding steps $\mathrm{d}u(\omega)$ and $\mathrm{d}v(\omega)$, which are random variables defined for $\omega \geq 0$, satisfy the following conditions:

$$\mathbb{E}\big[u(\omega)\big] = \mathbb{E}\big[v(\omega)\big] = 0 \quad \text{for } \omega \geq 0$$
$$\mathbb{E}\big[u^2(\omega)\big] = \mathbb{E}\big[v^2(\omega)\big] = 2S_{X_0}(\omega) \quad \text{for } \omega \geq 0$$
$$\mathbb{E}\big[u(\omega) \cdot v(\omega')\big] = 0 \quad \text{for } \omega, \ \omega' \geq 0$$
$$\mathbb{E}\big[\mathrm{d}u(\omega)\big] = \mathbb{E}\big[\mathrm{d}v(\omega)\big] = 0 \quad \text{for } \omega \geq 0 \qquad (2.40)$$
$$\mathbb{E}\big[\mathrm{d}u^2(\omega)\big] = \mathbb{E}\big[\mathrm{d}v^2(\omega)\big] = 2S_{X_0}(\omega) \quad \text{for } \omega \geq 0$$

$$\mathbb{E}\big[du(\omega)du(\omega')\big] = \mathbb{E}\big[dv(\omega)dv(\omega')\big] = 0 \quad \text{for } \omega, \ \omega' \geq 0, \omega \neq \omega'$$
$$\mathbb{E}\big[du(\omega)dv(\omega')\big] = 0 \quad \text{for } \omega, \ \omega' \geq 0$$

In the second equation of Eq. (2.40), $S_{X_0}(\omega)$ is the differential spectral density function, whose first derivative is the spectral density function $S_X(\omega)$

$$\frac{dS_{X_0}(\omega)}{d\omega} = S_X(\omega), \quad \text{for } \omega \geq 0 \tag{2.41}$$

The inequality $\omega \neq \omega'$ in the sixth relationship of Eq. (2.40) ensures that the frequency ranges $(\omega + d\omega)$ and $(\omega' + d\omega')$ do not overlap. The spectral representation of the stationary stochastic process of Eq. (2.39) has zero-mean value and autocorrelation function equal to the target $R_X(\tau)$ since

$$\mathbb{E}\big[X(t,\theta)\big] =$$
$$= \mathbb{E}\{\int_0^\infty \big[\cos(\omega t)du(\omega) + \sin(\omega t)dv(\omega)\big]\}$$
$$= \int_0^\infty \{\cos(\omega t)\mathbb{E}\big[du(\omega)\big] + \sin(\omega t)\mathbb{E}\big[dv(\omega)\big]\}$$
$$= 0 \tag{2.42}$$

The autocorrelation function can be expressed as

$$\mathbb{E}\big[X(t,\theta)X(t+\tau,\theta)\big] = \mathbb{E}\{\int_0^\infty \big[\cos(\omega t)du(\omega) + \sin(\omega t)dv(\omega)\big]$$
$$\ldots \int_0^\infty \big[\cos(\omega' \cdot (t+\tau))du(\omega') + \sin(\omega' \cdot (t+\tau))dv(\omega')\big]\}$$

$$\tag{2.43}$$

$$= \int_0^\infty \int_0^\infty \cos(\omega t) \cdot \cos(\omega'(t+\tau))\}\mathbb{E}\big[du(\omega)du(\omega')\big]+$$
$$\ldots + \int_0^\infty \int_0^\infty \sin(\omega t) \cdot \sin(\omega'\{t+\tau\})\}\mathbb{E}\big[du(\omega)dv(\omega')\big]+$$
$$\ldots + \int_0^\infty \int_0^\infty \sin(\omega t) \cdot \cos(\omega'\{t+\tau\})\}\mathbb{E}\big[dv(\omega)du(\omega')\big]+$$
$$\ldots + \int_0^\infty \int_0^\infty \sin(\omega t) \cdot \sin(\omega'\{t+\tau\})\}\mathbb{E}\big[dv(\omega)dv(\omega')\big]$$

Using the last three relations of Eq. (2.40) and the above relation for $\omega = \omega'$ together with the trigonometric equality $(\cos(a - b) = \cos a \cos b + \sin a \sin b)$ we get

$$\mathbb{E}\big[X(t,\theta)X(t+\tau,\theta)\big] = \int_0^\infty \cos(\omega t)\cos\{\omega(t+\tau)\}2S_X(\omega)d\omega + \ldots$$

$$+ \int_0^\infty \sin(\omega t)\sin\{\omega(t+\tau)\}2S_X(\omega)d\omega + \ldots$$

$$= \int_0^\infty \cos(\omega\tau)2S_X(\omega)d\omega$$

Because the power spectral density function is an even function and $\sin(\omega\tau)$ an odd one, it stands that

$$2\int_0^\infty S_X(\omega)d\omega = \int_{-\infty}^\infty S_X(\omega)d\omega \qquad (2.44)$$

$$\text{and} \quad \int_{-\infty}^\infty S_X(\omega)\sin(\omega\tau)d\omega = 0 \qquad (2.45)$$

Finally we get

$$\mathbb{E}\big[X(t,\theta)\big] = \int_{-\infty}^\infty S_X(\omega)\cos(\omega\tau)d\omega$$

$$= \int_{-\infty}^\infty S_X(\omega)e^{i\omega\tau}d\omega =$$

$$= R_X(\tau) \qquad (2.46)$$

Rewriting Eq. (2.39) in the following form:

$$X(t,\theta) = \sum_{k=0}^\infty \big[\cos(\omega_k t)du(\omega_k) + \sin(\omega_k t)dv(\omega_k)\big] \qquad (2.47)$$

where $\omega_k = k\Delta\omega$ and setting $du(\omega_k)$ and $dv(\omega_k)$ as

$$du(\omega_k) = X_k$$

$$dv(\omega_k) = Y_k \qquad (2.48)$$

and if X_k and Y_k are independent random variables with zero-mean value and standard deviation equal to $\sqrt{2S_X(\omega_k)\Delta\omega}$, it can be easily proven that Eq. (2.40) is satisfied. By replacing Eq. (2.48) to (2.47) we get

$$X(t,\theta) = \sum_{k=0}^\infty \big[\cos(\omega_k t)\cdot X_k + \sin(\omega_k t)Y_k\big] \qquad (2.49)$$

From the other hand, if we define $du(\omega_k)$ and $dv(\omega_k)$ as

$$du(\omega_k) = \sqrt{2}A_k \cos(\Phi_k)$$
$$dv(\omega_k) = -\sqrt{2}A_k \sin(\Phi_k) \tag{2.50}$$

where $A_k = \sqrt{2S_X(\omega_k)\Delta\omega}$ and Φ_k are independent random phase angles uniformly distributed over the range $[0, 2\pi]$, it can be easily demonstrated that the conditions of Eq. (2.40) are satisfied. Indeed we have

$$\mathbb{E}\big[du(\omega_k)\big] = \mathbb{E}\big[\sqrt{2}A_k \cos(\Phi_k)\big] =$$
$$= \sqrt{2}A_k \int_{-\infty}^{\infty} \cos(\Phi_k)p(\Phi_k)\mathrm{d}\Phi_k \tag{2.51}$$

where $p(\Phi_k)$ is the probability density function of the random variable Φ_k with type

$$p(\Phi_k) = \begin{cases} \frac{1}{2\pi} & \text{if } 0 \le \Phi_k \le 2\pi \\ 0 & \text{else} \end{cases}$$

By combing the last two equations we get

$$\mathbb{E}\big[du(\omega_k)\big] = \sqrt{2}A_k \int_0^{2\pi} \frac{1}{2\pi} \cos(\Phi_k)\mathrm{d}\Phi_k = 0 \tag{2.52}$$

In the same manner, we have $\mathbb{E}\big[dv(\omega_k)\big] = 0$. Consequently we calculate

$$\mathbb{E}\big[du^2(\omega_k)\big] = \mathbb{E}\big[A_k^2 \cos^2(\Phi_k)\big] =$$
$$= 2A_k^2 \int_0^{2\pi} \frac{1}{2\pi}\big(1 + \cos(\Phi_k)\big)\frac{1}{2\pi}\mathrm{d}\Phi_k$$
$$= 2A_k^2 \frac{1}{2\pi} = 2S_X(\omega) \tag{2.53}$$

$\mathbb{E}\big[dv^2(\omega_k)\big]$ is estimated in the same way. Finally, for the random field we get

$$X(t,\theta) = \sum_{k=0}^{\infty} \big[\cos(\omega_k t) \cdot \sqrt{2}\big(2S_X(\omega_k)\Delta\omega\big)^{\frac{1}{2}} \cos(\Phi_k) -$$
$$\ldots - \sin(\omega_k t)\sqrt{2}\big(2S_X(\omega_k)\Delta\omega\big)^{\frac{1}{2}} \sin(\Phi_k)\big] =$$
$$= \sqrt{2}\sum_{k=0}^{\infty} \big(2S_X(\omega_k)\Delta\omega\big)^{\frac{1}{2}} \cos(\omega_k t + \Phi_k) \tag{2.54}$$

2.2.3 Simulation Formula for Stationary Stochastic Fields

In order to have a realization of this $X(t, \theta)$ we need to truncate the summation of
Eq. (2.54) after N terms.

$$\hat{X}(t, \theta) = \sqrt{2} \sum_{n=0}^{N-1} A_n \cos(\omega_n t + \Phi_n) \tag{2.55}$$

where

$$A_n = \left(2 S_X(\omega_k) \Delta\omega\right)^{\frac{1}{2}} \quad \text{for } n = 0, 1, \ldots, N-1 \tag{2.56}$$

$$\omega_n = n \Delta\omega$$

$$\Delta\omega = \frac{\omega_u}{N}$$

$$A_0 = 0 \quad \text{or } S_X(\omega_0 = 0) = 0$$

The coefficient A_0 is chosen zero such that the temporal mean value averaged over the
whole simulation time $T_0 = \frac{2\pi}{\Delta\omega}$ of the generated stochastic process $\hat{X}(t, \theta)$ remains
zero in each generated sample. This is because if some power spectral contribution
is added at $\omega = 0$, a random variable term is always present, shifting the temporal
(sample) average apart from being zero. In order to avoid having to impose this
condition the frequency shifting theorem was proposed by Zerva (1992) but with the
side effect of doubling the period of the simulated field.

In Eq. (2.56) ω_u is usually applied as the uppercut off frequency after which the
power spectrum becomes practically zero. In order to estimate this frequency we use
the following criterion:

$$\int_0^{\omega_u} S_X(\omega) d\omega = (1 - \epsilon) \int_0^\infty S_X(\omega) d\omega \tag{2.57}$$

where $\epsilon \ll 1$ is the "admissible relative error". The target autocorrelation function
$R_{\hat{X}}(\tau)$ is given by

$$R_{\hat{X}}(\tau) = \int_{-\omega_u}^{\omega_u} S_X(\omega) e^{i\omega\tau} d\omega = \int_0^{\omega_u} 2 S_X(\omega) \cos \omega\tau \, d\omega \tag{2.58}$$

The difference between these two functions

$$\epsilon^*(\tau) = R_X(\tau) - R_{\hat{X}}(\tau) = \int_{\omega_u}^\infty 2 S_X(\omega) \cos(\omega\tau) d\omega \tag{2.59}$$

corresponds to the mean square simulation error due to the truncation of the spectral density function for $|\omega| \geqslant \omega_u$, which is termed "truncation error".

One sample function of the stochastic process can be generated by replacing the phase angles $\Phi_0, \ldots, \Phi_{N-1}$ in Eq. (2.55) with their corresponding sample values $\phi_0(\omega), \ldots, \phi_{N-1}(\omega)$, as these can be generated by some random number generator as follows:

$$\hat{X}(t, \theta) = \sqrt{2} \sum_{n=0}^{N-1} A_n \cos(\omega_n t + \phi_n(\theta)) \tag{2.60}$$

It must be mentioned that the step Δt of the generated sample functions must satisfy the following condition in order to avoid aliasing.

$$\Delta t \leq \frac{\pi}{\omega_u} \tag{2.61}$$

The sample functions generated by Eq. (2.60) are obviously bounded by

$$\left|\hat{X}(t, \theta)\right| \leq \sqrt{2} \sum_{n=0}^{N-1} A_n \tag{2.62}$$

For the cases of 2D and 3D spectral representation, Eq. (2.60) takes the form

$$\hat{X}(t, \theta) = \hat{X}(t_1, t_2, \theta) = \sqrt{2} \sum_{i=1}^{N_1} \sum_{j=1}^{N_2} A_{ij} [\cos(\omega_{1i} t_1 + \omega_{1j} t_2 + \phi_{ij}^1(\theta)) +$$
$$+ \cos(\omega_{1i} t_1 - \omega_{2j} t_2 + \phi_{ij}^2(\theta))]$$

and

$$\hat{X}(t, \theta) = \hat{X}(t_1, t_2, t_3; \theta) = \sqrt{2} \sum_{i=1}^{N_1} \sum_{j=1}^{N_2} \sum_{k=1}^{N_3} A_{ijk} [\cos(\omega_{1i} t_1 + \omega_{2j} t_2 + \omega_{3k} t_3 + \phi_{ijk}^1(\theta)) +$$
$$+ \cos(\omega_{1i} t_1 + \omega_{2j} t_2 - \omega_{3k} t_3 + \phi_{ijk}^2(\theta)) +$$
$$+ \cos(\omega_{1i} t_1 - \omega_{2j} t_2 + \omega_{3k} t_3 + \phi_{ijk}^3(\omega)) + \tag{2.63}$$
$$+ \cos(\omega_{1i} t_1 - \omega_{2j} t_2 - \omega_{3k} t_3 + \phi_{ijk}^4(\omega))]$$

respectively, with

$$A_{ij} = \sqrt{2 S_X(\omega_1, \omega_2) \Delta\omega_1 \Delta\omega_2}$$
$$A_{ijk} = \sqrt{2 S_X(\omega_1, \omega_2, \omega_3) \Delta\omega_1 \Delta\omega_2 \Delta\omega_3}$$

$$\Delta\omega_{1,2,3} = \frac{\omega_{(1,2,3)u}}{N_{1,2,3}}$$

$$\omega_{1,2,3} = (t_1, t_2, t_3)\Delta\omega_{1,2,3}$$

The numbers $N_{1,2}$ and $N_{1,2,3}$ of independent angle phases $\varphi(\theta)$ generated randomly in the range $[0, 2\pi]$ for the cases of two and three-dimensional random fields, respectively, are:

$$N_{1,2} = 2N_1 N_2 \tag{2.64}$$

and

$$N_{1,2,3} = 4N_1 N_2 N_3 \tag{2.65}$$

respectively.

2.3 Non-Gaussian Stochastic Processes

In nature, most of the uncertain quantities appearing in engineering systems are non-Gaussian (e.g., material, geometric properties, seismic loads). Nevertheless, the Gaussian assumption is often used due to lack of relevant experimental data and for simplicity in the mathematical implementation. It must be noted that this assumption can be problematic in many cases. For example, in the case where the Young's modulus is assumed to be a random variable following a Gaussian distribution, negative values for the Young's modulus may occur which have no physical meaning. For this reason, the problem of simulating non-Gaussian stochastic processes and fields has received considerable attention. However, the KL and Spectral representation methods, as discussed above, are limited in generating realizations of Gaussian stochastic processes due to the central-limit theorem, since the random variables in the summation formulas are independent.

In order to fully characterize a non-Gaussian stochastic process all the joint multidimensional density functions are needed which is generally not possible. For this reason a simple transformation of some underlying Gaussian field with known second-order statistics can be used in order to simulate a non-Gaussian stochastic process. If $X(t, \theta)$ is a stationary zero-mean Gaussian process with unit variance and spectral density function (SDF) $S_X(\omega)$, a homogeneous non-Gaussian stochastic process $y(t, \theta)$ with power spectrum $S_y^T(\omega)$ can be defined as

$$y(t, \theta) = Y^{-1}\Phi[X(t, \theta)] \tag{2.66}$$

where Φ is the standard Gaussian cumulative distribution function and Y is the non-Gaussian marginal cumulative distribution function of $y(t, \theta)$. The transformation $Y^{-1}\Phi$ is a memory-less translation since the value of $y(t, \theta)$ at an arbitrary point t depends on the value of $X(t, \theta)$ at the same point only. The resulting non-Gaussian field $y(t, \theta)$ is called a translation field.

The main shortcoming of translation fields is that, although the mapped sample functions of (2.66) will have the prescribed target marginal probability distribution Y, their power spectrum will not be identical to $S_y^T(\omega)$. Another important issue, pointed out by Grigoriu (1984), is that the choice of the marginal distribution of $y(t, \theta)$ imposes constraints to its correlation structure. In other words, Y and $S_y^T(\omega)$ have to satisfy a specific compatibility condition derived directly from the definition of the autocorrelation function of the translation field as

$$R_y^T(\tau) = \int_{-\infty}^{+\infty} \int_{-\infty}^{+\infty} Y^{-1}[\Phi(X_1)]Y^{-1}[\Phi(X_2)]\phi[X_1, X_2; R_X(\tau)]dX_1 dX_2 \quad (2.67)$$

where $X_1 = X(t, \theta)$, $X_2 = X(t + \tau, \theta)$ and ϕ is the pdf of the underlying Gaussian field. If these two quantities are proven to be incompatible through (2.67), then no translation field can be found having the prescribed characteristics. In this case, one has to resort to translation fields that match the target marginal distribution and/or SDF approximately.

2.4   Solved Numerical Examples

1. For an 1D-1V, zero-mean, Gaussian stochastic process $X(t, \theta)$ defined in the range $[-a, a]$ and with autocorrelation function $R_X(t_i, t_j)$ given by

$$R(t_i, t_j) = \sigma^2 \exp\left(-\frac{|t_i - t_j|}{b}\right) \quad (2.68)$$

Solve the Fredholm integral equation in order to estimate the eigenvalues and the eigenfunctions of the KL expansion.

Solution:

The Fredholm integral equation of the second kind is defined as

$$\int_{-a}^{+a} \sigma^2 \exp\left(-c|t_i - t_j|\right)\varphi(t_j)dt_j = \lambda \varphi(t_i) \quad (2.69)$$

where $c = 1/b$. Equation (2.69) can be written as

$$\int_{-a}^{t_i} \sigma^2 e^{\left(-c|t_i - t_j|\right)}\varphi(t_j)dt_j + \int_{t_i}^{+a} \sigma^2 e^{\left(-c|t_i - t_j|\right)}\varphi(t_j)dt_j = \lambda \varphi(t_i) \quad (2.70)$$

Differentiating the above equation once with respect to t_i gives

$$\lambda \varphi'(t_i) = -\sigma^2 c e^{-ct_i} \int_{-a}^{t_i} e^{ct_j}\varphi(t_j)dt_j + \sigma^2 c e^{ct_i} \int_{t_i}^{+a} \sigma^2 e^{-ct_j}\varphi(t_j)dt_j \quad (2.71)$$

and differentiating a second time gives

$$\lambda \varphi''(t) = (\lambda c^2 - 2\sigma^2 c)\varphi(t) \tag{2.72}$$

If we define $\omega^2 = \frac{-\lambda c^2 + 2\sigma^2 c}{\lambda}$ then Eq. (2.72) becomes

$$\varphi''(t) + \omega^2 \varphi(t) = 0 \tag{2.73}$$

Thus, the integral in Eq. (2.69) is transformed to the ordinary differential equations of Eq. (2.73) and evaluating Eqs. (2.70) and (2.71) at $t = -a$ and $t = a$ we can estimate its boundary conditions as

$$c\varphi(a) + \varphi'(a) = 0$$
$$c\varphi(-a) - \varphi'(-a) = 0 \tag{2.74}$$

Solving these equations simultaneously we get the eigenvalues and eigenfunctions described in Sect. 2.2.1.

2. Consider an 1D-1V truncated (its values are either bounded below and/or above), zero-mean Gaussian (TG) stochastic process $X(t, \theta)$, obtained from an underlying Gaussian process (denoted by $g(t, \theta)$), as follows:

$$X(t, \theta) = \begin{cases} g(t, \theta) & \text{if } |g(t, \theta)| \le 0.9 \\ 0.9 & \text{otherwise} \end{cases}$$

Use the following two spectral density functions for the description of $g(t, \theta)$

$$\text{SDF}_1 : S_g(\omega) = \frac{1}{4}\sigma_g^2 b^3 \omega^2 e^{-b|\omega|} \tag{2.75}$$

$$\text{SDF}_2 : S_g(\omega) = \frac{1}{2\pi}\sigma_g^2 \sqrt{\pi} b \omega^2 e^{-\frac{1}{4}b\omega^2} \tag{2.76}$$

with $\sigma = 0.2$ and $b = 1$. Generate realizations of the non-Gaussian process $X(t, \theta)$ and estimate its power spectral density $S_{X_{TG}}(\omega)$.

Solution:

Spectrum SDF_1 has zero power at $\omega = 0$, while spectrum SDF_2 has its maximum value at $\omega = 0$. For both spectra, b is a correlation length parameter. Simulate the underlying Gaussian process $g(t)$ according to SDF1 and SDF2 using the spectral representation method and then get the truncated Gaussian process $X_{TG}(t, \theta)$ by truncating the simulated Gaussian process $g(t, \theta)$ in the following way: if $g(t, \theta) > 0.9$ set $g(t, \theta) = 0.9$ or if $g(t, \theta) < -0.9$ set $g(t, \theta) = -0.9$.

Because the simulated non-Gaussian process $X_{TG}(t, \theta)$ is obtained as a nonlinear transformation of the underlying Gaussian process $g(t, \theta)$, its spectral density functions $S_{X_{TG}}(\omega)$ is going to be different from SDF$_1$ and SDF$_2$. The new spectral density functions of the truncated fields can be computed by producing samples $X_{TG}(t, \theta)$ and computing the spectra from Eq. (1.28). The $S_{X_{TG}}(\omega)$ is eventually determined by ensemble averaging.

2.5 Exercises

1. For a nonstationary Wiener–Levy stochastic process with covariance function

$$C_X(t_i, t_j) = \sigma^2 \min(t_i, t_j) \tag{2.77}$$

   where $0 \leq t \leq a$, the Fredholm integral equation can be solved analytically. Find the eigenvalues and eigenfunctions.
2. For an 1D-1V stationary and zero-mean Gaussian stochastic process $X(t, \theta)$ with $t \in [0, 1]$ the autocorrelation function $R_X(t_i, t_j)$ is defined as

$$R(t_i, t_j) = \sigma^2 \exp\left(\frac{|t_i - t_j|}{b}\right) \tag{2.78}$$

   with σ being the standard deviation and b the correlation length parameter. Using the KL expansion:

   a. Calculate the first 10 eigenvalues and eigenfunctions by solving the Fredholm integral equation analytically and numerically, for $\sigma = b = 1$.
   b. For $b = 1$ estimate the approximated covariance function and compare it with the exact one using $M = 2, 5, 10$ terms in the truncated KL.
   c. Plot 1000 realizations of the approximated process and calculate its probability distribution at $t = 0.5$.

3. For the 1D-1V zero-mean Gaussian stochastic process $X(t, \theta)$ with autocorrelation function

$$R_X(\tau) = \sigma^2 \frac{b^2(b - 2\tau)}{(b + 3)} \tag{2.79}$$

   where b is the correlation length parameter, σ^2 is the variance of the stochastic process and $\tau \in [0, 10]$:

   a. Generate 1000 realizations of the process for $b = 1$ and $\sigma^2 = 0.8$ using the spectral representation method.
   b. Discretize the domain in $n = 11$ points, i.e., 10 elements, and represent the stochastic field at these points using the midpoint method.

   c. Use the local average method to simulate the random variables at the same midpoints. Compare the properties of the random process with these of the midpoint method.

4. For the Gaussian stochastic process of exercise 3, simulate the 1000 realizations using the KL method for $M = 1, \ldots, 6$ terms in the truncated series, and $b = 1$ and 10. Estimate the corresponding power spectrums and compare them with the exact ones.
5. Generate 1000 realizations of (a) a lognormal translation process, and (b) a triangular pdf translation process using the underlying Gaussian process with the two spectral densities functions SDF_1, SDF_2 given in example 2 in the section of solved numerical examples and then estimate the resulting non-Gaussian Spectra.

Chapter 3
Stochastic Finite Element Method

3.1 Stochastic Principle of Virtual Work

The basis of the displacement-based finite element method (FEM) is the principle of virtual work. This principle states that the body in Fig. 3.1 which is in equilibrium under Dirichlet[1] boundary conditions in S_u and Neumann[2] in the remaining S_f domain (well-posed boundary value problem), requires that for any compatible small virtual displacements

$$\overline{\mathbf{u}}(\mathbf{x}) = [\overline{u}(\mathbf{x}), \overline{v}(\mathbf{x}), \overline{w}(\mathbf{x})] \tag{3.1}$$

imposed on the body in its state of equilibrium, the total internal virtual work is equal to the total external virtual work:

$$\delta \overline{W} = \overline{W}_{external} - \overline{W}_{internal} = 0, \tag{3.2}$$

where $\mathbf{x} = (x, y, z)$ is the position vector, $\overline{W}_{external}$ is the total virtual work of the external forces, and $\overline{W}_{internal}$ is the total virtual work of the internal forces. In the deterministic FEM method, the body is discretized in a number of nodes and elements (Fig. 3.1) and with the use of element shape functions $\mathbf{N}^{(e)}(\mathbf{x}) = [N_1, \ldots, N_n]$, the internal displacement field $\mathbf{u}^{(e)}(\mathbf{x})$ of element e can be expressed as a function of its nodal displacement $\mathbf{d}^{(e)} = [d_1, \ldots, d_n]$ as

$$\mathbf{u}^{(e)}(x, y, z) = \mathbf{N}^{(e)\mathsf{T}} \mathbf{d}^{(e)} \tag{3.3}$$

---

[1]The Dirichlet boundary condition specifies the values that the solution needs to take on along the boundary of the domain.

[2]Neumann boundary condition specifies the values that the derivative of the solution needs to take on the boundary of the domain.

© Springer International Publishing AG 2018
V. Papadopoulos and D.G. Giovanis, *Stochastic Finite Element Methods*,
Mathematical Engineering, https://doi.org/10.1007/978-3-319-64528-5_3

Fig. 3.1 Three-dimensional
body and a representative
three-dimensional finite
element

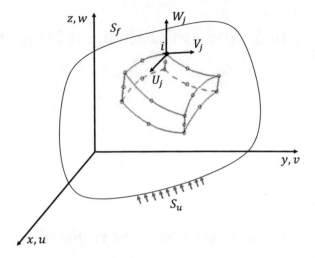

where n is the number of degrees of freedom of element e. The nodal strains are
given by:

$$\boldsymbol{\varepsilon}^{(e)} = \mathbf{B}^{(e)}\mathbf{d}^{(e)} \qquad (3.4)$$

where $\mathbf{B}$ is a matrix that relates the components of strains to the nodal displacements
and its elements are obtained by differentiation of the element shape functions with
respect to $\mathbf{x}$. Thus, the stresses can be expressed as:

$$\boldsymbol{\sigma}^{(e)} = \mathbf{D}\mathbf{B}^{(e)}\mathbf{d}^{(e)} \qquad (3.5)$$

where $\mathbf{D}$ is the matrix describing the constitutive properties of the material. In the
case that these properties are random, i.e., they vary in time and/or in space and their
variation can be described in the general case with a $n\mathrm{D} - m\mathrm{V}$ stochastic process,
the constitutive matrix $\mathbf{D}$ becomes random and can be written in the form:

$$\mathbf{D}(\mathbf{x}, \theta) = \mathbf{D}_0\big[1 + X(\mathbf{x}, \theta)\big] \qquad (3.6)$$

where $\mathbf{D}_0$ is the matrix containing the mean value of the properties and $X(\mathbf{x}, \theta)$ is
a zero-mean stationary stochastic process indicating the fluctuations around $\mathbf{D}_0$. By
enforcing to the element e, a virtual displacement field

$$\overline{\mathbf{u}}^{(e)} = \mathbf{N}^{(e)}\overline{\mathbf{d}}^{(e)} \qquad (3.7)$$

and introducing the nodal force vector $\mathbf{P}^{(e)}$, the principal of virtual work states that:

$$\overline{\mathbf{d}}^{(e)^T}\mathbf{P}^{(e)} = \int_{\Omega^e} \overline{\boldsymbol{\varepsilon}}^{(e)^T}\boldsymbol{\sigma}^{(e)}\mathrm{d}\Omega^e \qquad (3.8)$$

Substituting Eq. (3.6) to (3.7) we can write:

$$\overline{\mathbf{d}}^{(e)^\mathsf{T}} \mathbf{P}^{(e)} = \overline{\mathbf{d}}^{e^\mathsf{T}} \left[\int_{\Omega^e} \mathbf{B}^\mathsf{T} \mathbf{D}^e(\mathbf{x}, \theta) \mathbf{B}^e \mathrm{d}\Omega^e \right] \tag{3.9}$$

$$= \overline{\mathbf{d}}^{e^\mathsf{T}} \int_{\Omega^e} \left(\mathbf{B}^{e^\mathsf{T}} \mathbf{D}_0^e \mathbf{B}^e + \mathbf{B}^{e^\mathsf{T}} \mathbf{D}_0^e X(\mathbf{x}, \theta) \mathbf{B}^e \right) \mathrm{d}\Omega^{(e)}$$

The stiffness matrix associated with a given element of volume Ω_e reads:

$$\mathbf{k}^e = \int_{\Omega^e} \mathbf{B}^{e^\mathsf{T}} \mathbf{D}_0^e \mathbf{B}^e \mathrm{d}\Omega^e + \int_{\Omega^e} \mathbf{B}^{e^\mathsf{T}} \mathbf{D}_0^e \mathbf{B}^e X(\mathbf{x}, \theta) \mathrm{d}\Omega^e \tag{3.10}$$

or

$$\mathbf{k}^e = \mathbf{k}_0^e + \Delta \mathbf{k}^e \tag{3.11}$$

where $\mathbf{k}_0^e = \int_{\Omega^e} \mathbf{B}^{e^\mathsf{T}} \mathbf{D}_0^e \mathbf{B}^e \mathrm{d}\Omega^e$ and $\Delta \mathbf{k}^e = \int_{\Omega^e} \mathbf{B}^{e^\mathsf{T}} \mathbf{D}_0^e \mathbf{B}^e X(\mathbf{x}, \theta) \mathrm{d}\Omega^e$ are the deterministic and stochastic parts of the stochastic stiffness matrix, respectively. The global stochastic stiffness matrix can be expressed in the same manner as:

$$\mathbf{K} = \sum_{e}^{N_e} \mathbf{k}^e = \mathbf{K}_0 + \Delta \mathbf{K} \tag{3.12}$$

where N_e is the total number of elements. Stochastic finite element analysis is based on the solution of the following equation:

$$\mathbf{P} = (\mathbf{K}_0 + \Delta \mathbf{K}) \mathbf{U} \tag{3.13}$$

where $\mathbf{P}$ is the external force vector (which may also be random) and $\mathbf{U}$ the nodal displacement vector.

3.2 Nonintrusive Monte Carlo Simulation

The most commonly used method for the solution of Eq. (3.13) is the nonintrusive Monte Carlo approximation. According to this approach, N_{sim} samples of the stochastic stiffness matrix $\Delta \mathbf{K}$ are generated and the deterministic problem of Eq. (3.13) is solved N_{sim} times, generating a population for the response vector $\{\mathbf{U}_j\}$ with $j = 1, \ldots, N_{sim}$. The variance of $\mathbf{U}$ is estimated after a statistical post-process of the population. Let u_i be the response of the degree of freedom (i). Its mean value and standard deviation are given by

$$\mathbb{E}[u_i] = \frac{1}{N_{sim}} \sum_{j=1}^{N_{sim}} u_i(j) \tag{3.14}$$

$$\sigma^2(u_i) = \frac{1}{N_{sim} - 1} \sum_{j=1}^{N_{sim}} \left[u_i^2(j) - N_{sim}(\mathbb{E}[u_i])^2 \right] \tag{3.15}$$

In a similar manner, higher order moments of response quantities can be estimated and finally, the entire pdf can be constructed from the statistical post-process of the N_{sim} samples. More details in MCS methods and their convergence properties are presented in Chap. 4.

3.2.1   Neumann Series Expansion Method

Expanding the inverse of the stochastic stiffness matrix in a Neumann series, the solution of the static problem of Eq. (3.13) can be written as:

$$\mathbf{U} = (\mathbf{K}_0 + \Delta\mathbf{K})^{-1}\mathbf{P} = (\mathbf{I} + \mathbf{K}_0^{-1}\Delta\mathbf{K})^{-1}\mathbf{K}_0^{-1}P \tag{3.16}$$

where $\mathbf{I}$ is the identity matrix. Introducing $\mathbf{J} = \mathbf{K}_0^{-1}\Delta\mathbf{K}^{-1}$, the inverse of $(\mathbf{K}_0 + \Delta\mathbf{K})$ can be expanded in Neumann series as:

$$(\mathbf{K}_0 + \Delta\mathbf{K})^{-1} = (\mathbf{I} - \mathbf{J} + \mathbf{J}^2 - \mathbf{J}^3 + \cdots)\mathbf{K}_0^{-1} \tag{3.17}$$

$$= \sum_{k=0}^{\infty} (-\mathbf{K}_0^{-1}\Delta\mathbf{K})^k \mathbf{K}_0^{-1}$$

Consequently, the response can be expressed with the following series:

$$\mathbf{U} = (\mathbf{I} - \mathbf{J} + \mathbf{J}^2 - \mathbf{J}^3 + \cdots)\mathbf{U}_0 \tag{3.18}$$

or

$$\mathbf{U} = \mathbf{U}_0 - \mathbf{U}_1 + \mathbf{U}_2 - \mathbf{U}_3 + \cdots \tag{3.19}$$

The solution can also be expressed in the following recursive equation:

$$\mathbf{K}_0\mathbf{U}_i = \Delta\mathbf{K}\mathbf{U}_{i-1} \tag{3.20}$$

where $\mathbf{U}_0 = \mathbf{K}_0^{-1}\mathbf{P}$. The number of terms required in the series can be obtained with the following convergence criterion:

$$\frac{\|\mathbf{U}_i\|}{\|\sum_{k=0}^{i}(-1)^k\mathbf{U}_k\|} \leq \varepsilon_1 \tag{3.21}$$

where ε_1 is the convergence tolerance criterion which is checked after each calculation of an additional term in Eq. (3.19). The Neumann series converges if and only if all absolute eigenvalues of $\mathbf{J}$ are less than unit. This implies that convergence is guaranteed only if parameter variability is small (less than 20–30%).

3.2.2 The Weighted Integral Method

This method was developed in the early 90s and is claimed not to require any discretization of the stochastic field. Since the elements of matrix $\mathbf{B}$ are obtained by differentiation of the element shape functions, they are going to be polynomials on $(x; y; z)$ of some degree lower than the polynomial approximation of the displacements. Therefore, the stochastic part of the stiffness matrix in Eq. (3.11) can be rewritten as:

$$\Delta \mathbf{k}_{ij}^e(\theta) = \int_{\Omega_e} \mathbf{P}_{ij}(\mathbf{x}) X(\mathbf{x}, \theta) d\Omega_e, \tag{3.22}$$

where the coefficients of polynomial $\mathbf{P}_{ij}$ are obtained from those of matrices $\mathbf{B}$ and $\mathbf{D}$. We can write P_{ij} as:

$$\mathbf{P}_{ij}(\mathbf{x}) = \mathbf{P}_{ij}(x, y, z) = \sum_{l=1}^{N_w} a_{ij}^l x^{\alpha_l} y^{\beta_l} z^{\gamma_l} \tag{3.23}$$

where N_w is the number of monomials in $\mathbf{P}_{ij}$, each of them corresponding to a set of exponents $(\alpha_l, \beta_l, \gamma_l)$. Introducing the following weighted integrals for a random field $X(\mathbf{x}, \theta)$

$$\chi_l^e(\theta) = \int_{\Omega_e} x^{\alpha_l} y^{\beta_l} z^{\gamma_l} X(\mathbf{x}, \theta) d\Omega_e \tag{3.24}$$

it follows that

$$\Delta \mathbf{k}_{ij}^e(\theta) = \sum_{l=1}^{N_w} a_{ij}^l \chi_l^e(\theta) \tag{3.25}$$

Collecting now the coefficients a_{ij}^l in a matrix $\Delta \mathbf{k}_l^e$, the (stochastic) element stiffness matrix can be written as:

$$\mathbf{k}^e(\theta) = \mathbf{k}_0^e + \sum_{l=1}^{N_w} \Delta \mathbf{k}_l^e \chi_l^e(\theta) \tag{3.26}$$

In the above equation, $\mathbf{k}_0^e$ and $\Delta \mathbf{k}_l^e$, $l = 1, \ldots, N_w$ are deterministic matrices and χ_l^e are weighted integrals which are random variables. For example, a 2-node truss element has $N_w = 1$, a 2-node two-dimensional beam $N_w = 2$, and a 4-node plane stress quadrilateral element has $N_w = 3$. The stochastic part of the global stiffness matrix is given by

$$\mathbf{K}(\theta) = \sum_{e=1}^{N_e} \left(\mathbf{k}_0^e + \sum_{l=1}^{N_w} \Delta \mathbf{k}_l^e \chi_l^e(\theta) \right), \tag{3.27}$$

where N_e is the total number of finite elements.

3.3 Perturbation-Taylor Series Expansion Method

Perturbation-Taylor series expansion methods have been proposed as an alternative to the computationally expensive MCS. These methods are usually combined with point discretization methods (see Sect. 2.1) in which the $nD - mV$ stochastic process $X(t, \theta)$ is discretized in N zero-mean random variables $\{a_i\}_{i=1}^N$. Following this approach, the Taylor series expansion of the stochastic stiffness matrix around its mean value $\mathbf{K}_0$ is given by the following relation:

$$\mathbf{K} = \mathbf{K}_0 + \sum_{i=1}^N \mathbf{K}_i^I a_i + \frac{1}{2} \sum_{j=1}^N \sum_{i=1}^N \mathbf{K}_{ij}^{II} a_i a_j + \ldots, \tag{3.28}$$

where $\mathbf{K}_i^I$ denotes the first partial derivative of $\mathbf{K}$ to the random variable a_i and $\mathbf{K}_{ij}^{II}$ is the second partial derivative of $\mathbf{K}$ to the random variables a_i and a_j, evaluated at $\{a\} = 0$:

$$\mathbf{K}_i^I = \left. \frac{\partial \mathbf{K}}{\partial a_i} \right|_{\{a\}=0}$$

$$\mathbf{K}_{ij}^{II} = \left. \frac{\partial \mathbf{K}}{\partial a_i \partial a_j} \right|_{\{a\}=0} \tag{3.29}$$

In order to solve the static problem of Eq. (3.13), we need to expand the displacement vector $\mathbf{U}$ and the external forces vector $\mathbf{P}$ with a Taylor series around the random variables a_i as well:

$$\mathbf{U} = \mathbf{U}_0 + \sum_{i=1}^N \mathbf{U}_i^I a_i + \frac{1}{2} \sum_{j=1}^N \sum_{i=1}^N \mathbf{U}_{ij}^{II} a_i a_j + \ldots \tag{3.30}$$

$$\mathbf{P} = \mathbf{P}_0 + \sum_{i=1}^N \mathbf{P}_i^I a_i + \frac{1}{2} \sum_{j=1}^N \sum_{i=1}^N \mathbf{P}_{ij}^{II} a_i a_j + \ldots, \tag{3.31}$$

where $\mathbf{U}_0$ and $\mathbf{P}_0$ are the mean displacement and force vectors and $\mathbf{U}_i^I, \mathbf{P}_i^I, \mathbf{U}_{ij}^{II}, \mathbf{P}_{ij}^{II}$ their first and second derivatives with respect to the random variables a_i, a_j at $a_i = a_j = 0$. Thus, the displacement vector can be obtained using the following successive solution procedure:

$$\mathbf{U}_0 = \mathbf{K}_0^{-1}\mathbf{P}_0$$
$$\mathbf{U}_i^I = \mathbf{K}_0^{-1}(\mathbf{P}_i^{-1} - \mathbf{K}_i^I\mathbf{U}_0)$$
$$\mathbf{U}_{ij}^{II} = \mathbf{K}_0^{-1}(\mathbf{P}_{ij}^{-1} - \mathbf{K}_i^I\mathbf{U}_i^I - \mathbf{K}_j^I\mathbf{U}_i^I - \mathbf{K}_{ij}^{II}\mathbf{U}_0) \quad (3.32)$$

Taking now the statistics of the response when using a first-order approximation, the mean and the covariance can be computed in closed form as:

Mean value:
$$\mathbb{E}_1[\mathbf{U}] = U_0 \quad (3.33)$$

Covariance:

$$\text{COV}_1(\mathbf{U}, \mathbf{U}) = \mathbb{E}\left\{[\mathbf{U} - \mathbb{E}[\mathbf{U}]][\mathbf{U} - \mathbb{E}[\mathbf{U}]]^\mathsf{T}\right\}$$

$$= \sum_{i=1}^{N}\sum_{i=1}^{N}\mathbf{U}_i^I(\mathbf{U}_j^I)^\mathsf{T}\mathbb{E}[a_i a_j] \quad (3.34)$$

Variance:

$$\text{Var}(\mathbf{U}, \mathbf{U}) = \text{diag}\left[\text{COV}_1(\mathbf{U}, \mathbf{U})\right]$$

$$= \sum_{i=1}^{N}\sum_{i=1}^{N}\text{diag}\left[\mathbf{U}_i^I(\mathbf{U}_j^I)^\mathsf{T}\right]\mathbb{E}[a_i a_j] \quad (3.35)$$

where the subscript 1 denotes the first order.

In the case of Gaussian random variables, the second-order approximation of the mean value and the covariance are given by:

Mean value:
$$\mathbb{E}_2(u) = \mathbb{E}_1[\mathbf{U}] + \frac{1}{2}\sum_{i=1}^{N}\sum_{i=1}^{N}\mathbf{U}_{ij}^{II}\mathbb{E}[a_i a_j] \quad (3.36)$$

Covariance:

$$\text{COV}_2(\mathbf{U}, \mathbf{U}) = \text{COV}_1(\mathbf{U}, \mathbf{U}) \quad (3.37)$$

$$+ \frac{1}{4}\sum_{i=1}^{N}\sum_{i=1}^{N}\sum_{k=1}^{N}\sum_{l=1}^{N}\mathbf{U}_{kl}^{II}\left[\mathbb{E}[a_i a_k]\mathbb{E}[a_j a_l] + \mathbb{E}[a_i a_l]\mathbb{E}[a_j a_k]\right]$$

The generally recognized disadvantage of perturbation/Taylor expansion methods is that their fair accuracy is limited allowing for their efficient implementation only to problems with small nonlinearities and small variations of the stochastic properties.

3.4   Intrusive Spectral Stochastic Finite Element Method (SSFEM)

The spectral stochastic finite element method (SSFEM) was proposed by Ghanem and Spanos (1991). In SSFEM, the stochastic field describing the uncertain parameters is discretized using the truncated KL expansion while the response, which is also random, is projected to an appropriate polynomial basis.

Consider the stochastic equilibrium equation as given by Eq. (3.13) in which $\mathbf{K}_0$ is the deterministic and $\Delta \mathbf{K}$ is the stochastic part of the stiffness matrix. If the covariance function of the stochastic field describing some material parameter variation is known, then the KL expansion can be used for representing the uncertainty, after calculating the eigenvalues and eigenfunctions of the Fredholm integral equation involved (see Sect. 2.2.1). In this case, the stochastic part $\Delta \mathbf{K}$ can be decomposed into a series of stochastic matrices as

$$\Delta \mathbf{K} = \sum_{i=1}^{\infty} \mathbf{K}_i \xi_i(\theta), \tag{3.38}$$

where $\mathbf{K}_i$ are the deterministic stiffness matrices which are functions of the eigenvalues and eigenfunctions of the KL expansion and $\{\boldsymbol{\xi}(\theta)\}$ are uncorrelated random variables. More specifically, $\mathbf{K}_i$ matrices are estimated as:

$$\mathbf{K}_i = \sum_{j=0}^{N_e} \sqrt{\lambda_i} \int_{\Omega^e} \varphi_i(x, y, z) \mathbf{B}^{e^T} \mathbf{D}_0^e \mathbf{B}^e \, d\Omega^e, \quad i = 1, \ldots, M \tag{3.39}$$

where M is the truncation order of the KL. The key idea of SSFEM is to project the solution of Eq. (3.13) onto an appropriate basis, in a manner similar to KL, since U it is also random. However, the KL expansion cannot be used for this projection, since the covariance function of $\mathbf{U}$ it is not known. For this reason, an alternative procedure that circumvents this problem is required.

3.4.1   Homogeneous Chaos

If the covariance function of a second-order stochastic process is not known (as in the case of the solution of a stochastic partial differential equation), then it can be expressed as a spectral expansion in terms of suitable orthogonal eigenfunctions with weights associated with a particular density. Commonly used random basis for this projection is the Polynomial chaos (PC) basis (also known as Wiener chaos) which is a Hilbert basis[3] in the space of random variables consisting of multidimensional Hermite polynomials. The PC expansion of the displacement field $\mathbf{u}$ in Eq. (3.13) is given by:

$$\mathbf{u}(\mathbf{x}) = a_0(\mathbf{x})\Gamma_0 + \sum_{i_1=1}^{\infty} a_{i_1}(\mathbf{x})\Gamma_1(\xi_{i_1})$$

$$+ \sum_{i_1=1}^{\infty}\sum_{i_2=1}^{i_1} a_{i_1 i_2}(\mathbf{x})\Gamma_2(\xi_{i_1}, \xi_{i_2})$$

$$+ \sum_{i_1=1}^{\infty}\sum_{i_2=1}^{i_1}\sum_{i_3=1}^{i_2} a_{i_1 i_2 i_3}(\mathbf{x})\Gamma_3(\xi_{i_1}, \xi_{i_2}, \xi_{i_3}) + \ldots \tag{3.40}$$

where $\Gamma_p(\cdot)$ are successive polynomial chaoses of their arguments. Equation (3.40) can be recast as

$$\mathbf{U} = \sum_{j=0}^{\infty} \hat{a}_j \Psi_j\{\boldsymbol{\xi}\} \tag{3.41}$$

where there is one-to-one correspondence between the functionals $\Psi(\cdot)$ and $\Gamma(\cdot)$ as well as between the coefficients $\hat{a}_j$ and $a_{i_1 i_2 \ldots i_r}(\mathbf{x})$. Using $\langle \cdot \rangle \equiv \mathbb{E}[\cdot]$ the M-dimensional polynomial chaoses have the following properties:

$$\langle \Gamma_0 \rangle = 1, \quad \langle \Gamma_{k>0} \rangle = 0, \quad \langle \Gamma_i \Gamma_j \rangle = 0, \text{ for } i \neq j \tag{3.42}$$

Since the Hermite polynomials are orthogonal with respect to the Gaussian probability measure, the polynomial chaos of order n can be obtained as:

$$\Gamma_n(\xi_{i_1}, \xi_{i_2}, \ldots, \xi_{i_n}) = (-1)^n \frac{\partial^n}{\partial \xi_{i_1}, \xi_{i_2}, \ldots, \xi_{i_n}} e^{\frac{1}{2}\boldsymbol{\xi}^T \boldsymbol{\xi}} \tag{3.43}$$

For all practical purposes Eq. (3.41) is truncated after P terms

$$\mathbf{U} = \sum_{j=0}^{P} \hat{a}_j \Psi_j\{\boldsymbol{\xi}\} \tag{3.44}$$

The value of P depends on the number of terms M in the truncated KL expansion of the input and the highest polynomial order p of the homogeneous chaos used, and is determined by the following formula

---

[3]I.e. it spans a Hilbert space which is defined as a vector space possessing an inner product as measure. Named after David Hilbert (1862–1943).

Table 3.1 One-dimensional polynomial chaoses

j	p	j-th basis polynomial Ψ_j
0	p = 0	1
1	p = 1	ξ_1
2	p = 2	$\xi_1^2 - 1$
3	p = 3	$\xi_1^3 - 3\xi_1$
4	p = 4	$\xi_1^4 - 6\xi_1^2 + 3$

Table 3.2 Two-dimensional polynomial chaoses

j	p	j-th basis polynomial Ψ_j
0	p = 0	1
1	p = 1	ξ_1
2		ξ_2
3	p = 2	$\xi_1^2 - 1$
4		$\xi_1\xi_2$
5		$\xi_2^2 - 1$
6	p = 3	$\xi_1^3 - 3\xi_1$
7		$\xi_1^2\xi_2 - \xi_2$
8		$\xi_1\xi_2^2 - \xi_1$
9		$\xi_2^3 - 3\xi_2$
10	p = 4	$\xi_1^4 - 6\xi_1^2 + 3$
11		$\xi_1^3\xi_2 - 3\xi_1\xi_2$
12		$\xi_1^2\xi_2^2 - \xi_1^2 - \xi_2^2 + 1$
13		$\xi_1\xi_2^3 - 3\xi_1\xi_2$
14		$\xi_2^4 - 6\xi_2^2 + 3$

$$P = \frac{(M + p)!}{M!p!} - 1 \tag{3.45}$$

For example, for $M = 1$ and $p = 4$ we get $P = 5$ terms (Table 3.1) while, for $M = 2$ and $p = 4$ we get $P = 15$ terms (Table 3.2). Note that the truncation length M determines the number of uncorrelated random variables used for the description of the stochastic input. For Gaussian fields the KL expansion is a special case of the PC expansion with $p = 1$.

3.4.2 Galerkin Minimization

Substituting Eqs. (3.38) and (3.44) into (3.13) and denoting $\xi_0(\theta) \equiv 1$, we get:

$$\mathbf{P} = \left(\mathbf{K}_0 + \sum_{i=1}^{M} \mathbf{K}_i \xi_i(\theta)\right)\left(\sum_{j=0}^{P} \hat{a}_j \Psi_j(\xi)\right)$$

$$= \sum_{i=0}^{M} \sum_{j=0}^{P} \mathbf{K}_i \xi_i(\theta) \hat{a}_j \Psi_j(\xi) \tag{3.46}$$

As a result, the residual error in Eq. (3.46) due to the truncation of the KL after M terms and the PC after P terms reads:

$$\varepsilon_P = \sum_{i=0}^{M} \sum_{j=0}^{P} \mathbf{K}_i \hat{a}_j \xi_i(\theta) \Psi_j(\xi) - \mathbf{P} = 0 \tag{3.47}$$

The best approximation of the exact solution $\mathbf{U}$ in the space spanned by $\{\Psi_k(\xi)\}_{k=0}^{P}$ is obtained by minimizing this residual in a mean square sense. This is equivalent to requiring this residual to be orthogonal to $\{\Psi_k(\xi)\}_{k=0}^{P}$, which yields:

$$\langle \varepsilon_P, \Psi_k \rangle = 0, \quad k = 0, \dots, P \tag{3.48}$$

By the orthogonality of $\{\Psi_k\}$, this reduces to a system of coupled deterministic equations from the solution of which we can estimate the coefficients of the truncated PC. By introducing the following notation:

$$c_{ijk} = \mathbb{E}[\xi_i \Psi_j \Psi_k] \tag{3.49}$$

$$\mathbf{P}_k = \mathbb{E}[\mathbf{P}\Psi_k] \tag{3.50}$$

Equation (3.46) can be rewritten as

$$\sum_{i=0}^{M} \sum_{j=0}^{P} c_{ijk} \mathbf{K}_i \hat{a}_j = \mathbf{P}_k, \quad k = 0, \dots, P \tag{3.51}$$

and by defining for the sake of simplicity

$$\mathbf{K}_{jk} = \sum_{i=0}^{M} c_{ijk} \mathbf{K}_i \tag{3.52}$$

Equation (3.51) rewrites as follows:

$$\sum_{j=0}^{P} \hat{a}_j \mathbf{K}_{jk} = \mathbf{P}_k \tag{3.53}$$

Fig. 3.2 Sparsity pattern of
K for the Gaussian case
($M = 2$, $p = 2$)

$$
\begin{bmatrix}
\mathbf{K}_0 & \mathbf{K}_1 & \mathbf{K}_2 & 0 & 0 & 0 \\
\mathbf{K}_1 & \mathbf{K}_0 & 0 & 2\mathbf{K}_1 & \mathbf{K}_2 & 0 \\
\mathbf{K}_2 & 0 & \mathbf{K}_0 & 0 & \mathbf{K}_1 & 2\mathbf{K}_2 \\
0 & 2\mathbf{K}_1 & 0 & 2\mathbf{K}_0 & 0 & 0 \\
0 & \mathbf{K}_2 & \mathbf{K}_1 & 0 & \mathbf{K}_0 & 0 \\
0 & 0 & 2\mathbf{K}_2 & 0 & 0 & 2\mathbf{K}_0
\end{bmatrix}
$$

In the above equation, each $\hat{a}_j$ is a N-dimensional vector and each $\mathbf{K}_{jk}$ is a matrix of size $N \times N$, where N is the size of the deterministic problem. The $P + 1$ different equations can be cast in a matrix form of size $N(P + 1) \times N(P + 1)$ as follows:

$$\mathbb{K}\mathbb{U} = \mathbb{P} \tag{3.54}$$

where

$$
\mathbb{K} =
\begin{bmatrix}
\sum_{i=0}^{M} c_{i00}\mathbf{K}_i & \sum_{i=0}^{M} c_{i10}\mathbf{K}_i & \cdots & \sum_{i=0}^{M} c_{iP0}\mathbf{K}_i \\
\sum_{i=0}^{M} c_{i01}\mathbf{K}_i & \sum_{i=0}^{M} c_{i11}\mathbf{K}_i & \cdots & \sum_{i=0}^{M} c_{iP1}\mathbf{K}_i \\
\vdots & \vdots & \ddots & \vdots \\
\sum_{i=0}^{M} c_{i0P}\mathbf{K}_i & \sum_{i=0}^{M} c_{i1P}\mathbf{K}_i & \cdots & \sum_{i=0}^{M} c_{iPP}\mathbf{K}_i
\end{bmatrix}
$$

and

$$\mathbb{U} = \begin{bmatrix} \mathbf{U}_0, \mathbf{U}_1, \cdots, \mathbf{U}_P \end{bmatrix}^{\mathsf{T}}$$

$$\mathbb{P} = \begin{bmatrix} \mathbf{P}_0, \mathbf{P}_1, \cdots, \mathbf{P}_P \end{bmatrix}^{\mathsf{T}}$$

The sparsity patterns of the ($N \times N$) nonzero block sub-matrices of **K** for a Gaussian input with $M = 2$ and $p = 2$ as well as $M = 6$ and $p = 4$ are shown in Figs. 3.2 and 3.3, respectively.

For large-scale problems, the solution of the augmented algebraic system of Eq. (3.54) can become quite challenging due to the increased memory and computational resources required. Furthermore, since the coefficients do not provide a clear interpretation of the response randomness by themselves, the following useful quantities are readily obtained:

Fig. 3.3 Sparsity pattern of **K** for the Gaussian case ($M = 6,\ p = 4$)

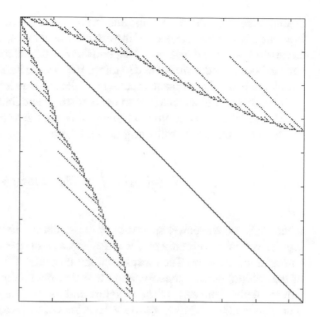

$$\mathbb{E}[\mathbf{U}] = \mathbb{E}\left[\sum_{j=0}^{P} \hat{a}_k \Psi_j\right] = \hat{a}_0 \mathbb{E}[\Psi_0] + \sum_{j=1}^{P} \hat{a}_j \mathbb{E}[\Psi_j] = \hat{a}_0 \qquad (3.55)$$

and

$$\mathrm{Var}(\mathbf{U}) = \mathbb{E}[(\hat{a} - \mathbb{E}[\hat{a}])^2] = \mathbb{E}\left[\left(\sum_{j=0}^{P} \hat{a}_j \Psi_j - \hat{a}_0\right)^2\right]$$

$$= \mathbb{E}\left[\left(\sum_{j=1}^{P} \hat{a}_j \Psi_j\right)^2\right] = \sum_{j=1}^{P} \hat{a}_j^2 \mathbb{E}[\Psi_k^2] \qquad (3.56)$$

Having calculated the coefficients $\hat{a}_j$, we can approximate the pdf of **U** from Eq. (3.41) by sampling from the distribution of $\boldsymbol{\xi}$ and plugging the samples into the PC. This step is performed at the end of SSFEM procedure at a minimum computational cost.

3.5 Closed Forms and Analytical Solutions with Variability Response Functions (VRFs)

The concept of variability response function (VRF) was introduced by Shinozuka in the late 1980s as an alternative to MCS, in order to provide information of the spatial distribution of the variability of a structure's response when no information of the spatial correlation structure of the uncertain system parameters is available. The VRF is assumed to be independent of the distributional and spectral characteristics of the uncertain system parameters and depend only on the deterministic system parameters such as boundary conditions and loading. VRFs were initially proposed for relatively simple statically determinant structures, while subsequent developments allowed for their successful implementation to general stochastic finite element systems.

VRF is defined as an integral expression for the variance of the response of a stochastic system in the following general form

$$\text{Var}[\mathbf{u}(\mathbf{x})] = \int_{-\infty}^{\infty} \text{VRF}(\mathbf{x}; \kappa) S_f(\kappa) d\kappa, \tag{3.57}$$

where $S_f(\kappa)$ is the power spectrum of the stochastic field $X(\mathbf{x}, \theta) \equiv f(\mathbf{x})$ describing the system uncertainty and $\kappa = (\kappa_x, \kappa_y, \kappa_z)$ are the wave numbers in the three Cartesian dimensions. The most important property of the VRF is its independence of the spectral density function and the pdf of the random field. It only depends on deterministic parameters of the structure and its boundary conditions (i.e., displacement and loading). Thus, once the VRF is known for a specific structure, performing only a simple integration gives the variance of the desired response quantity. The VRF provides a "spectral distribution-free" upper and lower bound on the variance of the response. As will be shown in the following, given a variance of the input stochasticity, a spectral density function defined by the delta function at the peak of the VRF provides the supremum of the response variance.

The major drawback of this method is that it can effectively provide with up to second-order information of the stochastic system response. Thus, if higher order moments of the marginal pdf of the response is required, one must resort to a classical intrusive SSFEM or nonintrusive MCS approach.

3.5.1   Exact VRF for Statically Determinate Beams

Although the stiffness formulation is commonly used in stochastic mechanics, in some cases it is advantageous to consider the compliance in the equilibrium equations. In the VRF generalism, this alternative is preferable because, as it will be shown next, it linearizes the expression for the response with respect to the uncertain parameter. For example, if the compliance of an uncertain material property corresponds to its bending rigidity, the homogeneous random field model for the flexibility can be written as:

$$\frac{1}{\text{EI}(x)} = \frac{1}{\text{E}_0\text{I}}(1 + f(x, \theta)) \tag{3.58}$$

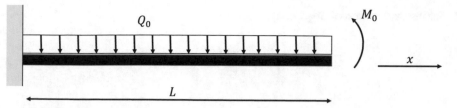

Fig. 3.4 Configuration of statically determinate beam

where $1/EI_0$ is the average flexibility and $f(x, \theta)$ is a zero-mean, 1D-1V homogeneous random field describing the fluctuations about its mean value. Consider the statically indeterminate beam of length L shown in Fig. 3.4, with a deterministic uniformly distributed load $q(x)$, a concentrated moment M_0 imposed at the free end and heterogeneous, random flexibility, $1/EI(x)$ defined in Eq. (3.58).

The governing differential equation for the displacement field, $u(x)$, for an Euler–Bernoulli beam is

$$\frac{d^2}{dx^2}\left(EI(x)\frac{du^2}{dx^2}\right) = q(x) \tag{3.59}$$

For a statically determinate beam, the moment distribution $M(x)$ is only dependent on the boundary conditions and loading, therefore it is deterministic and the differential equation reduces to:

$$\frac{d^2u}{dx^2} = \frac{M(x)}{EI(x)} \tag{3.60}$$

This allows $u(x)$ to be described by a Green's function independent of $M(x)$ and $EI(x)$ as:

$$u(x) = \int_0^x \frac{1}{EI(\xi)} G(x, \xi) M(\xi) d\xi$$

$$= \int_0^x \frac{(1 + f(\xi))}{EI_0} G(x, \xi) M(\xi) d\xi, \tag{3.61}$$

where $G(x, \xi)$ and $M(x)$ are the Green and the bending moment functions, respectively given by

$$G(x, \xi) = (x - \xi) \tag{3.62}$$

$$M(x) = -\frac{Q_0}{2}(L - x)^2 + M_0 \tag{3.63}$$

Taking expectation of displacement gives

$$
\mathbb{E}[u(x)] = \int_0^x \frac{(1 + \mathbb{E}[f(\xi)])}{\mathrm{EI}_0} G(x, \xi) M(\xi) \mathrm{d}\xi
$$

$$
= \int_0^x \frac{1}{\mathrm{EI}_0} G(x, \xi) M(\xi) \mathrm{d}\xi \tag{3.64}
$$

which is equal to the corresponding deterministic solution, and the mean square $\mathbb{E}[u(x)^2]$ is given by

$$
\frac{1}{(\mathrm{EI}_0)^2} \int_0^x \int_0^x \mathbb{E}[(1 + f(\xi_1))(1 + f(\xi_2))] G(x, \xi_1), G(x, \xi_2) M(\xi_1) M(\xi_2) \mathrm{d}\xi_1 \mathrm{d}\xi_2
$$

$$
= \frac{1}{(\mathrm{EI}_0)^2} \int_0^x \int_0^x (1 + \mathbb{E}[f(\xi_1) f(\xi_2)]) G(x, \xi_1), G(x, \xi_2) M(\xi_1) M(\xi_2) \mathrm{d}\xi_1 \mathrm{d}\xi_2
$$

$$
\tag{3.65}
$$

The response variance is then given by

$$
\mathrm{Var}[u(x)] = \mathbb{E}[u(x)^2] - (\mathbb{E}[u(x)])^2
$$

$$
= \frac{1}{(\mathrm{EI}_0)^2} \int_0^x \int_0^x R_f(\xi_1 - \xi_2) G(x, \xi_1), G(x, \xi_2) M(\xi_1) M(\xi_2) \mathrm{d}\xi_1 \mathrm{d}\xi_2
$$

$$
\tag{3.66}
$$

where $R_f(\xi_1 - \xi_2)$ denotes the autocorrelation function of the random field $f(x, \theta)$. Applying the Wiener–Khinchin transformation to the autocorrelation function in Eq. (3.66), the variance of the response displacement can be written as

$$
\mathrm{Var}[u(x)] = \int_{-\infty}^{\infty} \mathrm{VRF}(x, \kappa) S_f(\kappa) \mathrm{d}\kappa \tag{3.67}
$$

where the VRF is given by

$$
\mathrm{VRF}(x, \kappa) = \left| \frac{1}{\mathrm{EI}_0} \int_0^x G(x, \xi) M(\xi) e^{i\kappa\xi} \mathrm{d}\xi \right|^2 \tag{3.68}
$$

and $S_f(\kappa)$ denotes the power spectral density function of $f(x, \theta)$. The above expressions for the $\mathrm{Var}[u(x)]$ and the VRF are exact analytic expressions, since no approximations were made for their derivation. In addition, the expression for the VRF in Eq. (3.68) is a general one and can be applied to any statically determinate beam with any kind of boundary or loading conditions, using the appropriate Green and bending moment functions. Since both $S_f(\kappa)$ and VRF(x, κ) are even functions of κ, the variance of the displacement can be also written as

$$\text{Var}[u(x)] = 2 \int_0^\infty \text{VRF}(x, \kappa) S_f(\kappa) d\kappa \leq \text{VRF}(x, \kappa^{\text{max}}) \sigma_f^2 \qquad (3.69)$$

where κ^{max} is the wave number at which the VRF takes its maximum value, and σ_f^2 is the variance of stochastic field $f(x, \theta)$. The upper bound given in Eq. (3.69) is physically realizable since the form of $f(x, \theta)$ that produces it is known. Specifically, the variance of $u(x)$ attains its upper bound value of VRF$(x, \kappa^{\text{max}}) \sigma_f^2$ when $f(x, \theta)$ becomes a random sinusoid, i.e.

$$f(x, \theta) = \sqrt{2} \sigma_f \cos(\kappa^{\text{max}} x + \phi) \qquad (3.70)$$

In Eq. (3.70), ϕ is a random phase angle uniformly distributed in the range $[0, 2\pi]$. In this case, the corresponding spectral density function of $f(x, \theta)$ is a delta function at wave number κ^{max}

$$S_f(\kappa) = \sigma_f^2 \delta(\kappa - \kappa^{\text{max}}) \qquad (3.71)$$

while its marginal pdf is a U-beta probability distribution function given by

$$p_f(s) = \frac{1}{\pi \sqrt{2\sigma_f^2 - s^2}} \quad \text{with} \quad -\sqrt{2}\sigma_f \leq s \leq \sqrt{2}\sigma_f \qquad (3.72)$$

The upper bound provided in Eq. (3.69) is spectral- and probability-distribution-free, since the only probabilistic parameter it depends on is the standard deviation of the inverse of the elastic modulus.

3.5.2   VRF Approximation for General Stochastic FEM Systems

A rigorous proof of the existence of the VRF and the derivation of its exact analytical form is available only for statically determinate systems. However, in further developments of this approach by Papadopoulos et al. (2006), the existence of a similar to Eq. (3.57) closed-form integral expression for the variance of the response displacement of the following form

$$\text{Var}[\mathbf{u}(\mathbf{x})] = \int_{-\infty}^{\infty} \text{VRF}(\mathbf{x}, \kappa; \sigma_f) S_f(\kappa) \mathrm{d}\kappa \qquad (3.73)$$

was demonstrated for general linear stochastic FEM systems under static loads using again flexibility-based formulation. For such general systems, VRF depends on standard deviation σ_f but appears to be independent of the functional form of the spectral density function $S_f(\kappa)$ modeling the inverse of the elastic modulus. The existence however of this integral expression had to be conjectured. Further investigations by Deodatis and Miranda (2012) verified the aforementioned results but showed that VRF has a slight dependence on the marginal pdf of the stochastic field modeling flexibility.

3.5.3 Fast Monte Carlo Simulation

For the case of general stochastic FEM systems for which closed form expressions are not available, the variability response function can be estimated numerically using a fast Monte Carlo simulation (FMCS) approach whose basic idea is to consider the stochastic field as a random sinusoid.

Steps of the FMCS Approach

Without loss of generality, for an 1D-1V stochastic field $f(x, \theta)$ the FMCS is implemented as follows:

1. For every wave number $\bar{\kappa}$, generate N sample functions of the stochastic field $f(x, \theta)$ as a random sinusoid

$$f_j(x, \theta) = \sqrt{2}\sigma_f \cos(\bar{\kappa}x + \phi_j(\theta)), \quad j = 1, \ldots, N \qquad (3.74)$$

   where σ_f is the standard deviation of the stochastic field and $\phi(\theta)$ are random phase angles uniformly distributed in the range $[0, 2\pi]$. Rather than picking up ϕ randomly in $[0, 2\pi]$, they can be selected at N equal intervals in $[0, 2\pi]$ for significant computational savings.[4]
2. Using these N generated sample functions of $f_j(x, \theta)$, it is straightforward to compute the corresponding N displacement responses either analytically or numerically. Then, the mean value of the response $\text{Var}(x)_{\bar{\kappa}}$ and its variance $\text{Var}[u(x)]_{\bar{\kappa}}$ for the specific value of $\bar{\kappa}$ can be easily determined by ensemble averaging the N computed responses.
3. The value of the VRF at some point x, for wave number $\bar{\kappa}$ and for standard deviation σ_f is computed from

---

[4]This is a simple 1D implementation of the Latin Hypercube Sampling methodology.

$$\text{VRF}(x; \overline{\kappa}, \sigma_f) = \frac{\text{Var}[u(x)]_{\overline{\kappa}}}{\sigma_f^2} \qquad (3.75)$$

4. Steps 1–3 are repeated for different values of the wave number $\overline{\kappa}$ of the random sinusoid. Consequently, $\text{VRF}(x; \overline{\kappa}, \sigma_f)$ is computed over a wide range of wave numbers, wave number by wave number. The entire procedure can be eventually repeated for different values of the standard deviation σ_f.

It should be pointed out that the FMCS can be implemented into the framework of a deterministic finite element code making this approach very general.

3.5.4 Extension to Two-Dimensional FEM Problems

The proposed methodology can be extended to two-dimensional problems in a straightforward manner. The inverse of the elastic modulus is now assumed to vary randomly over a 2D domain according to the following equation:

$$\frac{1}{E(x, y)} = F_0(1 + f(x, y)) \qquad (3.76)$$

where E is the elastic modulus, F_0 is the mean value of the inverse of E, and $f(x, y)$ is now a 2D-1V, zero-mean homogeneous stochastic field modeling the variation of $1/E$ around its mean value F_0. Accordingly, the integral expressions for the variance of the displacement $u(x, y)$ become:

$$\text{Var}[u(x, y)] = \int_{-\infty}^{\infty} \int_{-\infty}^{\infty} \text{VRF}(x, \kappa_x; y, \kappa_y; \sigma_f) S_f(\kappa_x, \kappa_y) d\kappa_x d\kappa_y \qquad (3.77)$$

where $\text{VRF}(x, \kappa_x; y, \kappa_y; \sigma_f)$ is the two-dimensional versions of VRF possessing the following biquadrant symmetry:

$$\text{VRF}(\kappa_x, \kappa_y) = \text{VRF}(-\kappa_x, -\kappa_y) \qquad (3.78)$$

$S_f(\kappa_x, \kappa_y)$ is the spectral density function of stochastic field $f(x, y)$ possessing the same symmetry.

The FEM-FMCS procedure described earlier for 1D beam problems can be used for 2D problems also, in order to estimate the VRF. The 1D random sinusoid in Eq. (3.74) now becomes a 2D one with the following form:

$$f_j(x, y; \theta) = \sqrt{2}\sigma_f \cos(\overline{\kappa}_x x + \overline{\kappa}_y y + \phi_j(\theta)), \quad j = 1, \ldots, N \qquad (3.79)$$

while the 2D VRF can be estimated via

$$\text{VRF}(x, \overline{\kappa}_x; y, \overline{\kappa}_y; \sigma_f) = \frac{\text{Var}[u(x, y)]_{\overline{\kappa}=\kappa_x, \overline{\kappa}=\kappa_y}}{\sigma_f^2} \qquad (3.80)$$

Upper bounds on the mean and variance of the response displacement can be established for the 2D case as follows:

$$\text{Var}[u(x, y)] \leq \text{VRF}(x, \kappa_x^{\max}; y, \kappa_y^{\max}; \sigma_f)\sigma_f^2 \qquad (3.81)$$

where $\kappa_x^{\max}, \kappa_y^{\max}$ is the wave number pair at which VRF takes its maximum value.

3.6 Solved Numerical Examples

1. Consider the cantilever beam of Fig. 3.4 with a squared cross section $0.5 \times 0.5\,\text{m}^2$, $L = 10\,\text{m}$, loaded with a concentrated force $P = 10\,\text{kN}$ at its free end. The Young modulus is described by a homogeneous stochastic field as $E(x) = E_0(1 + f(x))$ where $E_0 = 3.0 \times 10^7\,\text{kN/m}^2$ and $f(x)$ a zero-mean homogeneous field.
   Calculate the mean and variance of the tip vertical displacement using 1000 MCS and the midpoint method, for the case that $f(x)$ is a random sinusoid of Eq. (3.74) for $\overline{k} = 2\pi/10$. Perform a convergence study with respect to the number of elements used for the FEM discretization of the cantilever.

   Solution:

   Figure 3.5a–c presents an exact sample realization of the random sinusoid for $\overline{k} = 2\pi/10$ along with the $f(x)$ representation, using the midpoint method for (a) $N = 1$, (b) $N = 2$ and (c) $N = 4$ elements.
   From this figure, it can be deduced that using only one element along the period of the field misrepresents the quality of the field. Using two elements improves the image of the field but lacks significantly accuracy. At least four elements should be used for a fair accuracy of the field representation. This result is also verified by the convergence study presented in Fig. 3.5d, where the variance of the response displacement computed after 1000 Monte Carlo Simulations is plotted as a function of the number of elements used for the FEM discretization. This leads to the general conclusion that the finite element mesh size should be at least less than $0.25T_u$ where T_u is the period corresponding to the upper cut frequency of the power spectrum.

2. Consider again the statically determinate beam of Fig. 3.4 with $L = 10\,\text{m}$, $Q_0 = 1\,\text{kN/m}$ and $M_0 = 10\,\text{kNm}$. The inverse of the elastic modulus is assumed to vary randomly according to Eq. (3.58) with $\frac{1}{E_0 I} = 8 \times 10^{-8}\,(\text{kNm}^2)^{-1}$. Calculate and plot the VRF and compute an upper bound for the variance of the displacement at the right end of the beam.

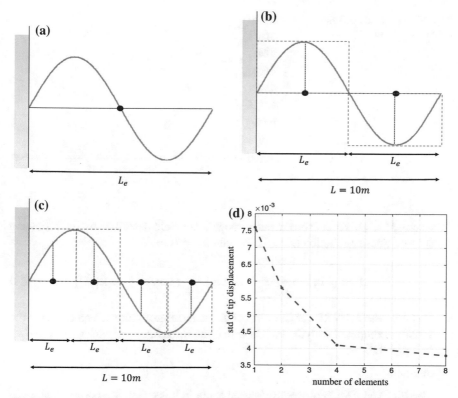

Fig. 3.5 Representation of the stochastic field using the midpoint method and **a** $N = 1$ element, **b** $N = 2$ elements, **c** $N = 4$ and **d** convergence of the variance of tip displacement as a function of the number of finite elements

Solution:

A plot of the VRF calculated at $x = L$ via Eq. (3.57) is presented in Fig. 3.6. From this figure, it can be seen that VRF is a smooth function of κ which tends to zero as κ goes to infinity. In addition, for this particular κ, VRF is maximized at $\overline{\kappa}^{max} = 0.4$. According to Eq. (3.69), an upperbound for the variance can be determined as $Var[u(x)] < 0.0038\sigma_f^2$.

3. Consider the cylindrical panel shown in Fig. 3.7. The nodes on the curved edge of the panel are assumed to be free to move in all directions, while the nodes along the straight sides are assumed to be hinged (fixed against translation). Dimensions and material properties are provided in Fig. 3.7. The loading consists of a vertical force equal to $P = 1414\,\text{kN}$ applied at the apex of the shell (point A in Fig. 3.7). The inverse of the elastic modulus is assumed to vary randomly over the surface of the unfolded panel according to:

$$\frac{1}{E(x, y)} = F_0(1 + f(x, y)), \tag{3.82}$$

Fig. 3.6 VRF of statistically
determinate beam

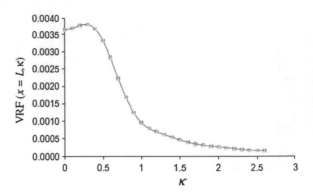

where $f(x, y)$ is a 2D zero mean homogeneous field described by the following
spectral density function (Eq. (3.83)) with $b_x = b_y = 2.0$:

$$S_f(\kappa_x, \kappa_y) = \frac{1}{4\pi}\sigma_g^2 b_x b_y \exp\{-\frac{1}{4}[(b_x\kappa_x)^2 + (b_y\kappa_y)^2]\} \qquad (3.83)$$

Compute the 2D VRF for the vertical displacement of point A and compare the
variance computed with the FMCS with brute force MCS for various levels of σ_f
in Eq. (3.83).

Solution:

The 2D FEM-FMCS procedure is used again in this example in order to numerically estimate the 2D VRF function of this shell-type structure. For this purpose,
the structure is analyzed for each wave number pair in Eq. (3.80) using the FEM
method in which the structure is discretized to 200 shell finite elements. Denoting
u_{A_v} the vertical displacement of point A, and VRF_{A_v} the corresponding variability response function, Fig. 3.8 provides plot of VRF_{A_v} calculated for a standard
deviation $\sigma_f = 0.4$, with max $VRF_{A_v} = 1.15 \times 10^{-4}$ at $\kappa_x = \kappa_y = 0$.
Figure 3.9 provides a plot of $\mathrm{Var}[u_{A_v}]$, as a function of σ_f computed with brute
force Monte Carlo as well as with the integral expression in Eq. (3.77). The values
for variance of the response displacement computed using the integral form are
very close to the corresponding values computed through brute force Monte Carlo
simulations, regardless of the value of the standard deviation used for modeling
the inverse of the modulus of elasticity.

3.7 Exercises

1. For the rod of Fig. 3.10 with length $L = 4\,\mathrm{m}$ and with a circular section of radius
   $R = 2.5\,\mathrm{cm}$ we apply a deterministic horizontal load $P = 100\,\mathrm{kN}$ at its most right
   edge.

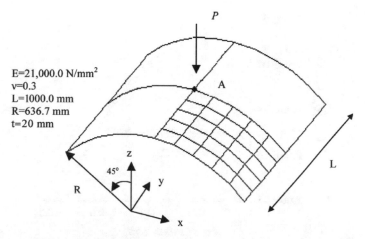

Fig. 3.7 Geometry, loading, finite element mesh, and material properties of the cylindrical panel. Note that although not shown, every little "rectangle" in the above figure is subdivided into two "triangular" finite elements

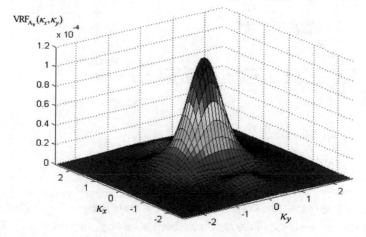

Fig. 3.8 Variability response function calculated using the 2D FEM-FMCS approach for the cylindrical panel shown in Fig. 3.7 and $\sigma_f = 0.4$

The inverse of the Young's modulus is considered to be a homogeneous random field model given by

$$\frac{1}{E(x)} = F_0(1 + f(x)) \tag{3.84}$$

where $F_0 = (1.25 \times 10^8 \text{ kN/m})^{-1}$ is the mean value of the inverse of $E(x)$ and $f(x)$ is a homogeneous, zero-mean stochastic field with the power spectrum of Eq. (2.75) for $b = 1$ and $\sigma_g = 0.1$ variance $\sigma_f^2 = 0.1$. The rod is discretized to

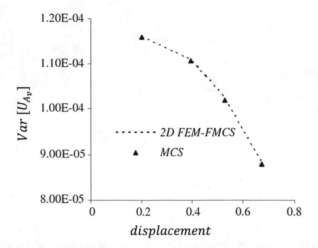

Fig. 3.9 Variance of response displacement u_{A_v}: comparison of results using Eq. (3.77) and from brute force Monte Carlo simulations (MCS). Plots correspond to SDF$_3$ with $b_x = b_y = 2.0$. The corresponding PDF is a truncated Gaussian

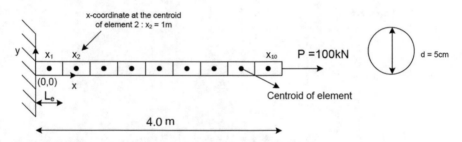

Fig. 3.10 Clumped rod with stochastic properties

10 truss finite elements. (1) Generate $N = 10^4$ realizations of $\frac{1}{E_{(x)}}$, (2) Calculate the mean and variance of the maximum horizontal displacement (u_{max}) of the rod using the local average as well as the midpoint method, (3) Estimate and then plot the pdf and cdf of u_{max}, (4) Compare the results for the two methods.

2. Compute analytically the VRF of the statically determinate cantilever of example 2 of Sect. 3.6. Compute again this VRF with the FMCS and compare the results.

3. Compute the deterministic matrices of the weighted integral method for a 2D truss finite element. Implement this result to the truss problem of exercise 2.

4. For the same truss compute the VRF analytically and compare the results with the weighted integral method as well as with brute force MCS.

5. Compute the augmented stiffness matrix of the previous truss example using SSFEM with $M = 2$ and $p = 2$ and compare the results of the SSFEM analysis with brute force MCS.

6. Compute the upper bounds of the response variance of the vertical displacement of point A in the solved numerical example 3.

Chapter 4
Reliability Analysis

There are different levels of reliability analysis, which can be used in any design methodology depending on the importance of the structure. The term "level" is characterized by the extent of information about the problem that is used and provided. The methods of safety analysis proposed currently can be grouped under four basic "levels" (namely **Levels I, II, III, IV**) depending upon the degree of sophistication applied to the treatment of the various problems.

In **Level I** methods, the probabilistic aspect of the problem is taken into account by introducing into the safety analysis suitable "characteristic values" of the random variables, conceived as fractile of a predefined order of the statistical distributions concerned. These characteristic values are associated with partial safety factors that should be deduced from probabilistic considerations so as to ensure appropriate levels of reliability in the design. In this method, the reliability of the design deviates from the target value, and the objective is to minimize such an error.

As **Level II** are classified, the reliability methods which employ up to second-order information of the uncertain parameter (i.e., mean and variance), supplemented with a measure of the correlation between parameters.

Level III methods encompass complete probabilistic analysis of the problem and involve the integration of the multidimensional joint probability density function of the random variables extended over some safety domain.

Level IV methods are appropriate for structures that are of major economic importance, involve the principles of engineering economic analysis under uncertainty, and consider costs and benefits of construction, maintenance, repair, consequences of failure, interest on capital, etc. Foundations for sensitive projects like nuclear power projects, transmission towers, highway bridges, are suitable objects of level IV design. The inherent probabilistic nature of design parameters, material properties and loading conditions involved in structural analysis is an important factor that influences structural safety. Reliability analysis leads to safety measures that a design engineer has to take into account due to the aforementioned uncertainties.

© Springer International Publishing AG 2018   71
V. Papadopoulos and D.G. Giovanis, *Stochastic Finite Element Methods*,
Mathematical Engineering, https://doi.org/10.1007/978-3-319-64528-5_4

4.1 Definition

The reliability L of a structure is defined as complement of the probability of failure and represents the probability of not failing:

$$L = 1 - P_f \tag{4.1}$$

where P_f is the probability of failure of the structure. In the case of time-independent reliability analysis where the design parameters are independent of time and independent between them, the probability of failure can be expressed as follows:

$$P_f = P(R - S \leq 0) \tag{4.2}$$

wherein the limit-state criterion is defined in terms of the some action S and some resistance R, each of which is described by a known probability density function $f_S(s)$, $f_R(r)$, respectively. Extending Eq. (4.2), the probability of failure P_f can be expressed as follows:

$$P_f = P(R \leq S) = P(R \leq x|_{S=x}) = \sum_x [P(R \leq x|_{S=x}) P(S = x)]$$

$$\Rightarrow P_f = \int_{-\infty}^{\infty} F_R(x) f_S(x) \mathrm{d}x \tag{4.3}$$

where $F_R(\cdot)$ is the cumulative distribution function of R:

$$F_R(x) = \int_{-\infty}^{x} f_R(x) \mathrm{d}x \tag{4.4}$$

Thus, by replacing Eq. (4.4) in (4.3) we get

$$P_f = \int_{-\infty}^{\infty} \int_{-\infty}^{S} f_R(R) f_S(S) \mathrm{d}R \mathrm{d}S \tag{4.5}$$

Quite general density functions f_R and f_S for R and S, respectively, are shown in Fig. 4.1 in which we can see the geometric interpretation of the probability of failure. For a given value x of the action S, the probability of failure is equal to the product of the black shaded surface, whose area is equal to $F_S(x)$, multiplied by the crosshatch surface which represents the value $f_S(x)\mathrm{d}x$. The probability of failure is equal to the sum of these products of the surfaces for all x.

Defining now a failure function $G(R - S) = R - S$ in the space of the variables S and R, termed as **limit-state function** with negative values defining the failure scenario, then Eq. (4.5) can be alternatively written as follows:

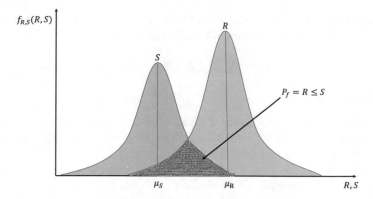

Fig. 4.1 Schematic representation of the probability, the resistance R to be smaller than the value x of the action S

$$P_f = \int_{G(R-S)\leq 0} f_{RS}(R, S) \mathrm{d}R\mathrm{d}S \qquad (4.6)$$

where f_{RS} is the joint probability density function of the random variables R, S. In the case where the random variables are independent:

$$P_f = \int_{G(R-S)\leq 0} f_R(R) f_S(S) \mathrm{d}R\mathrm{d}S \qquad (4.7)$$

In the general case where a large number of dependent random variables are involved in the structural problem, the failure function is expressed as $G(\mathbf{X})$ where $\mathbf{X} = [X_1, \ldots, X_n]$ is the vector of the basic random variables, and Eq. (4.6) is rewritten as:

$$P_f = \int_{G(\mathbf{X})\leq 0} f_\mathbf{X}(\mathbf{X}) \mathrm{d}\mathbf{X} \qquad (4.8)$$

where $f_\mathbf{X}(\cdot)$ is the joint distribution function. In cases of a large number of random variables and/or strongly nonlinear failure function $G(\mathbf{X})$ the calculation of this integral becomes extremely difficult.

4.1.1 Linear Limit-State Functions

The most widely used probability distribution function for the description of the uncertain system parameters in reliability analysis of structures is the Gaussian. By definition the joint probability distribution function of the random vector $\mathbf{X} = [X_1, \ldots, X_n]$ is given by Eq. (B.46) as:

$$f_\mathbf{X}(\mathbf{X}) = \frac{1}{(2\pi)^{n/2}|\mathbf{C}|^{1/2}} \exp\left(-\frac{1}{2}(\mathbf{X} - \boldsymbol{\mu})^\mathsf{T}\mathbf{M}(\mathbf{X} - \boldsymbol{\mu}) \right) \qquad (4.9)$$

where μ is the mean value of each random variable, C is the covariance matrix and $M = C^{-1}$. If the random variables are independent then

$$f_X(X) = f_{X_1}(X_1) f_{X_2}(X_2) \cdots \cdots f_{X_n}(X_n) \tag{4.10}$$

Fundamental Case

The fundamental case in reliability analysis is when the random variables are two, namely the R and S, are independent. In addition, they follow the normal distribution with mean μ_R and μ_S and variance σ_R^2 and σ_S^2, respectively, then the limit-state function $Z = R - S$ has a mean and variance given by well-known rules for addition of normal random variables:

$$\mu_Z = \mu_R - \mu_S, \quad \sigma_Z^2 = \sqrt{\sigma_R^2 + \sigma_S^2} \tag{4.11}$$

By definition the probability of failure is expressed as follows:

$$P_f = P(Z \le 0) = P\left(\frac{Z - \mu_Z}{\sigma_Z} \le 0 - \frac{-\mu_Z}{\sigma_Z}\right) = \Phi\left(\frac{0 - \mu_Z}{\sigma_Z}\right) \tag{4.12}$$

The random variable $Z' = \frac{Z - \mu_Z}{\sigma_Z}$ is the standard normal variable with mean value equal to zero and unit variance. Thus, by using Eqs. (4.11) and (4.12) we get the probability of failure as follows:

$$P_f = \Phi\left[\frac{-(\mu_R - \mu_S)}{(\sigma_R^2 + \sigma_S^2)^{1/2}}\right] = \Phi(-\beta) \tag{4.13}$$

where $\beta = \mu_Z/\sigma_Z$ is defined as reliability (safety) index. If either of the standard deviations σ_R and σ_S or both are increased, the term in square brackets in (4.13) will become smaller and hence P_f will increase. Similarly, if the difference between the mean of the load effect and the mean of the resistance is reduced, P_f increases.

General Case

Limit-state functions are in general nonlinear. However, it is possible to approximate the area near the failure point with a hyperplane, expressed in the space of the basic random variables. In this case the approximation is made with a **First Order Approximation Method** (FORM). Assuming that the failure function is expressed by the following linear relation

$$Z = a_0 + a_1 X_1 + \cdots + a_n X_n \tag{4.14}$$

the mean value can be estimated with the linear combination

$$\mu_Z = a_0 + a_1 \mu_{X_1} + \cdots + a_n \mu_{X_n} \tag{4.15}$$

The variance for the case of independent random variables is given by

$$\sigma_Z^2 = a_1^2 \sigma_{X_1}^2 + \cdots + a_n^2 \sigma_{X_n}^2 \tag{4.16}$$

while for dependent random variables is

$$\sigma_Z^2 = a_1^2 \sigma_{X_1}^2 + \cdots + a_n^2 \sigma_{X_n}^2 + \sum_{i \neq j} \rho(X_i, X_j) a_i a_j \sigma_{X_i} \sigma_{X_j} \tag{4.17}$$

where $\rho(X_i, X_j)$ is the correlation coefficient of X_i and X_j. Rewritting previous equations in a matrix form we have

$$\boldsymbol{\mu}_Z = a_0 + \mathbf{a}^T \boldsymbol{\mu}_X \tag{4.18}$$

$$\sigma_Z^2 = \mathbf{a}^T \mathbf{C} \mathbf{a} \tag{4.19}$$

where $\mathbf{a} = [a_1, \ldots, a_n]$ and $\mathbf{C}$ the covariance matrix. The calculation of the failure probability is then a straight forward calculation using Eq. (4.13).

Geometric Interpretation of the Reliability Index

The reliability index β, defined for both the fundamental as well as for the general case of linear limit-state functions, has the following geometric interpretation: In the fundamental case with two independent random variables S and R, their corresponding standard normal variables are given by the following:

$$R' = \frac{R - \mu_R}{\sigma_R}, \quad S' = \frac{S - \mu_S}{\sigma_S} \tag{4.20}$$

The failure function $G(R, S)$ is a straight line which in the standard normal space is rewritten as follows:

$$G(R, S) = R - S = \sigma_R R' - \sigma_S S' + \mu_R - \mu_S = G'(R', S') = 0 \tag{4.21}$$

With reference to Fig. 4.2 we can define β as the smallest distance $|OA|$ from the origin O to the line (or generally the hyperplane) forming the boundary between the safe domain and the failure domain, i.e., the domain defined by the failure event.

4.1.2 Nonlinear Limit-State Functions

The reliability index β can be defined in an analogous manner for the case of nonlinear failure functions in n-dimensional random spaces. In this case it is called the **Hasofer**

Fig. 4.2 Geometric interpretation of the safety index β - linear failure function

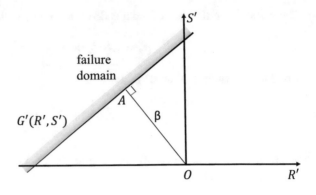

Fig. 4.3 Geometric interpretation of the safety index β - nonlinear failure function

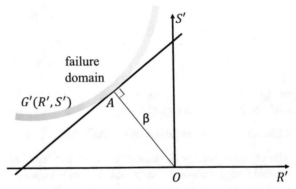

and Lind reliability index and it is defined as the smallest distance between the origin in the standard normal space and the tangent of the failure function (Fig. 4.3).

In the general case where the basic random variables are represented by vector $\mathbf{X} = [X_1, \ldots, X_n]$ with their corresponding standard normal variables $\mathbf{Z} = [Z_1, \ldots, Z_n]$, the vector $\overline{OA}$ of Fig. 4.3 can be expressed as follows:

$$\overline{OA} = \beta \mathbf{a} \tag{4.22}$$

where $\mathbf{a} = [a_1, \ldots, a_n]$ is the unit vector in the standard normal space for the direction which is vertical to the failure function at point A. The point A belongs to the failure function so we have the following:

$$G'(\beta a_1, \beta a_2, \ldots, \beta a_n) = 0 \tag{4.23}$$

Also, the vector $\overline{OA}$ is vertical to the tangent of the failure function G' in point A. This leads to the following relation:

$$\overline{\nabla G'}|_A = \left(\frac{\partial G'}{\partial Z_1}, \ldots, \frac{\partial G'}{\partial Z_n} \right)|_A = -k\overline{OA} = -k(a_1, \ldots, a_n) \Rightarrow$$

$$\Rightarrow a_i = -\frac{1}{k}\frac{\partial f}{\partial Z_1}, \quad i = 1, \ldots, n \tag{4.24}$$

and the coefficient k can be expressed as follows:

$$k = \frac{|\overline{\nabla G'}|_A}{|\mathbf{a}|} = |\overline{\nabla G'}|_A = \left(\sum_{i=1}^{n} \left[\frac{\partial G'}{\partial Z_1} (\beta a_i) \right]^2 \right)^{1/2} \tag{4.25}$$

since the vector $\mathbf{a}$ is a unit vector. Equations (4.24) and (4.25) are $n + 1$ equations that can be solved repeatively for the values of a_i and β. The point A is called **design point** and defines the point which is the most likely for failure to occur.

4.1.3   First- and Second-Order Approximation Methods

The geometric interpretation of the design point as the minimum distance between the origin and the failure function in the standard normal space can be interpreted into the constrained optimization problem of finding the minimum of the following:

$$d^2 = \mathbf{Z}\mathbf{Z}^T \tag{4.26}$$

under the restriction

$$G'(\mathbf{Z}) = 0 \tag{4.27}$$

The above optimization problem can be solved using standard gradient-based algorithms. These algorithms require the computation of the sensitivity of the failure function on the basic random variables, defined as follows:

$$|\overline{\nabla G'}| = \left(\frac{\partial G'}{\partial Z_1}, \ldots, \frac{\partial G'}{\partial Z_1} \right) \tag{4.28}$$

Therefore, the existence of a differential failure function is a prerequisite in order to calculate the derivatives of Eq. (4.28). As the analytical differentiation of the failure function $G(\mathbf{X})$ is, in most of the realistic structural reliability problems, intractable, the use of semi-analytical methods such as finite differences is necessary for an estimate of these derivatives. This is due to the fact that the failure equation of a structure consists of all possible failure mechanisms that can be defined for it as well as their combinations (system reliability). On the other hand, it is feasible to calculate points that belong in the failure function using finite element analysis and then have a first- (**FORM**) or second-order (**SORM**) interpolation of the failure domain in these points in order to estimate the derivatives of Eq. (4.28). In **FORM** methods, the failure function is approximated with a hyperplane around the design point while in **SORM** with a paraboloid. If the number of design parameters is n then **FORM** in conjunction with the finite difference method requires $2n$ finite element analysis, while **SORM** requires $2(n - 1)$ additional points in order to describe the parabola around the design point. Summarizing, the **FORM** and **SORM** methods can be very effective for simple structures with a small number of basic random variables. In more complex structures with a large number of random variables Monte Carlo simulation methods are found to be more capable of handling these problems.

4.2 Monte Carlo Simulation (MCS)

Monte Carlo simulation (MCS) is a technique that tries to solve a model by using random numbers. The term "Monte Carlo" was introduced by **Von Neumann**[1] and **Ulam**[2] during the second world war as a code name for the stochastic simulations that were conducted in order to build atomic bombs in Los Alamos laboratories. Since then the MCS has been used for the solution of multidimensional integrals in abstract domains, integral equations, and differential systems when the analytical maths fail to give a solution. In reliability analysis of structures, MCS is used for the solution of the integral in Eq. (4.8) and is particularly applicable when an analytical solution is not attainable and/or the failure domain cannot be expressed or approximated by an analytical form. This is mainly the case in problems of complex nature with a large number of basic variables.

4.2.1   The Law of Large Numbers

The performance of MCS is based on the **law of large numbers** which is associated with the behavior of the sum of a large number of random variables that follow any distribution function. If n is the number of random variables r_i with uniform distribution function in the range $[a, b]$. Then an unbiased estimate of an integral in $[a, b]$ can be computed as follows:

$$\frac{1}{b-a} \int_a^b f(r)\mathrm{d}r \approx \overline{F} = \frac{1}{n} \sum_{i=1}^n f(r_i) \quad \text{for } n \to \infty \qquad (4.29)$$

For large numbers of n, the estimator $\overline{F}$ is converging to the correct result of the integral. It must be mentioned that convergence here has a statistical definition which is different from the corresponding algebraical one. The definition of the statistical convergence is as follows:

A series $A(n)$ converges to number B as n approaches infinity, then and only then for every probability P, with $0 < P < 1$ and any positive number δ, there exists a natural number k such that for every $n > k$ the probability that $A(n)$ is in the range $[B - \delta, B + \delta]$ is P^n.

---

[1] John Von Neumann, Hungarian-American mathematician, physicist (1903–1957).
[2] Stanislav Marcin Ulam, Polish-American mathematician (1909–1984).

This definition of the statistical convergence is **weak** because regardless the size of n, $A(n)$ is not certain to be around B. To sum up, the rule of large number proves that the estimation of an integral with MCS is correct for an infinite number of simulations while the central limit theorem described in Sect. B.5 provides the information about the distribution of the estimator.

4.2.2 Random Number Generators

The success of the MCS is hidden in the way the random variables are generated and since we are talking about continuous variables, the quality of the generation depends on the following two parameters:

1. The quality of the uniform random number generator in the range [0, 1] generator.
2. The transformation of the uniformly distributed in [0, 1] random variables into continuous random variables with specific probability distribution function.

Regarding the first parameter, what is important is how good is the pseudorandom number generator (PRNG), i.e., what is the period of the algorithms generating the pseudorandom numbers. This period expressed, as the number of generated random numbers before repetition starts, must be much larger than the size of the required sample. The most famous methods for generating pseudorandom numbers are the ones that are based on a recursive algorithm that generates a sequence of numbers by the residual of the division of a linear transformation, with an integer number m. Obviously, this procedure is deterministic since every generated term is already known through the relation from which it is produced. The congruential methods are generally based on the following relation:

$$X_{i+1} = (aX_i + c)\mathrm{mod}m \qquad (4.30)$$

where a, c, and m are nonnegative integers. Providing Eq. (4.30) with an initial seed X_0 and dividing by m we get the normal distribution.

$$U_i = \frac{X_i}{m} \qquad (4.31)$$

It is obvious that the number will start repeating after m steps, so in order to guarantee a large period the value of m must be sufficiently large and the period length will be equal to m. Because this procedure is deterministic, the first time a number is repeated, the entire sequence is repeated. The algorithm of Eq. (4.30) has period m only if:

- c and m do not have a common divisor
- $a = 1(\mathrm{mod}g)$ where g is the first factor of m
- $a = 1(\mathrm{mod}4)$ if m is a multiple of 4.

For a computer using binary system the selection of $m = 2^\beta$, where β is the word length of the computer is the most appropriate choice. The number 2^β is the largest integer that can be saved in a computer. The parameter c must be an odd number and $a = 1[\text{mod}4] = 2^r + 1$ with $r \geq 2$. If $c = 0$ then we have a multiplicative pseudo-random number generator. In this case the maximum period is $m/4$ and is obtained when:

- The seed X_0 is odd.
- $a = 8r \pm 3, r \geq 0$

Regarding the transformation of this uniform number generation into random numbers with some continuous probability distribution, this is either based on the central limit theorem or in the inverse transform described in Sect. B.3.1.

4.2.3 Crude Monte Carlo Simulation

Crude Monte Carlo simulation was initially used as a hit or miss method in order to approximate one-dimensional integrals of the form

$$I = \int_a^b f(x)\mathrm{d}x \tag{4.32}$$

where $0 \leq f(x) \leq c$ and $a \leq x \leq b$. Let Ω be the domain defined by the following:

$$\Omega = \{(x, y) : a \leq x \leq b, 0 \leq y \leq c\} \tag{4.33}$$

If (x, y) is a random vector whose coordinates are uniformly distributed over the domain Ω, with probability distribution

$$f_{XY}(x, y) = \begin{cases} \frac{1}{c(b-a)} & \text{for } (x, y) \in \Omega \\ 0 & \text{otherwise} \end{cases}$$

then, as shown in Fig. 4.4, the probability p that the random vector lies under $f(x)$ is:

$$p = \frac{\text{area under } f(x)}{\Omega} = \frac{\int_a^b f(x)\mathrm{d}x}{c(b-a)} = \frac{I}{c(b-a)} \tag{4.34}$$

According to the rule of large numbers if we consider N random vectors (x_i, y_i) with $i = 1, \ldots, N$ then the probability p can be approximated by

$$\overline{p} = \frac{N_H}{N} \tag{4.35}$$

Fig. 4.4 Schematic representation of the hit and miss method

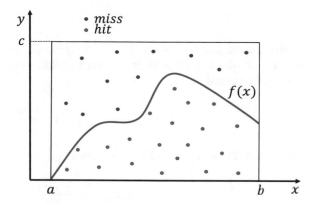

where N_H is the number of the events for which $f(x_i) \geq y_i$ (hit), N is the total number of simulations and $\overline{p}$ is the estimation of the probability while the integral of Eq. (4.32) can be estimated as follows:

$$\overline{I} = c(b-a)\overline{p} \tag{4.36}$$

Every trial of the N trials is a Bernoulli trial with mean value and standard deviation those of the dynamic distribution. Thus, the expectation of the estimator is given by

$$\mathbb{E}[\overline{I}] = c(b-a)\frac{\mathbb{E}[N_H]}{N} = pc(b-a) = I \tag{4.37}$$

which states that the expectation of the estimator $\overline{I}$ is equal to the true value of the integral I. Thus, $\overline{I}$ is an unbiased estimator of I. The variance of the estimator $\overline{p}$ is given by the following:

$$\mathrm{Var}(\overline{p}) = \frac{\mathrm{Var}(N_H)}{N} = \frac{1}{N^2}\mathrm{Var}(N_H) = \frac{p}{N}(1-p) \tag{4.38}$$

Equations (4.38) and (4.34) combined give

$$\mathrm{Var}(\overline{p}) = \frac{1}{N}\frac{1}{[c(b-a)]^2}[c(b-a)-I] \tag{4.39}$$

while Eqs. (4.36) and (4.38) give

$$\mathrm{Var}(\overline{I}) = \frac{1}{N}[c(b-a)-I] \tag{4.40}$$

One major question is how large must the number N be in order to have statistical convergence of the form

$$P[(\overline{I} - 1) < \varepsilon] \geq \alpha \tag{4.41}$$

where α is the desired tolerance, i.e., the range in which the estimator must be found, and ε is the desired tolerance in probability. Following the **Chebyshev inequality**[3] we obtain the following:

$$P[(\overline{I} - 1) < \varepsilon] \geq 1 - \frac{\mathrm{Var}(\overline{I})}{\varepsilon^2} \tag{4.42}$$

which in combination with Eq. (4.41) gives

$$\alpha \leq 1 - \frac{\mathrm{Var}(\overline{I})}{\varepsilon^2} \tag{4.43}$$

and we finally get

$$N \geq \frac{p(1 - p)[c(b - a)]^2}{(1 - \alpha)\varepsilon^2} \tag{4.44}$$

If MCS is pursued to estimate a probability in the order of 10^{-6} it must be expected that approximately 10^8 simulations are necessary to achieve an estimate with a coefficient of variance in the order of 10%. A large number of simulations are thus required using Monte Carlo simulation and all refinements of this crude technique have the purpose of reducing the variance of the estimate.

Despite the fact that the mathematical formulation of the MCS is relatively simple and the method has the capability of handling practically every possible case regardless of its complexity, this approach has not received an overwhelming acceptance due to the excessive computational effort that is required. Several sampling techniques, called variance reduction techniques, have been developed in order to improve the computational efficiency of MCS by reducing its statistical error and keeping the sample size to the minimum possible.

4.3   Variance Reduction Methods

Variance reduction techniques target in the utilization of every known information about the problem under consideration. If we have no information about the problem then these methods cannot be used. A way to obtain some information about the problem is to make an initial simulation, whose results can be used in order to define certain parameters to be used for reducing the variance of the subsequent simulation.

---

[3]Named after Pafnuty Chebyshev, Russian mathematician (1821–1894).

4.3.1   *Importance Sampling*

Importance sampling has been the most widely used variance reduction technique because of its simplicity and effectiveness. This technique is based on the idea of concentrating the distribution of the sampling points in regions of the sample space that are "more important" than others over the random space. Given the n-dimension integral of Eq. (4.32) the idea of importance sampling is to draw the sample from a proposal distribution and re-weight the integral by using importance weights targeting the correct distribution. Mathematically, this corresponds to a change of the integrated variables as follows:

$$f(x)\mathrm{d}x \rightarrow \frac{f(x)}{g(x)} g(x)\mathrm{d}x = q(x)\mathrm{d}z \tag{4.45}$$

where

$$q(x) = \frac{f(x)}{g(x)} \quad \text{and } \mathrm{d}z = q(x)\mathrm{d}x \tag{4.46}$$

$g(x)$ is called **importance sampling** function. Thus, the integral of Eq. (4.32) becomes

$$I = \int_a^b q(x)\mathrm{d}z = \mathbb{E}[q(x)] = \mathbb{E}\left[\frac{f(x)}{g(x)}\right] \tag{4.47}$$

and the estimator of Eq. (4.47) can be written as follows:

$$I \approx P = \frac{1}{N} \sum_{i=1}^N \left[\frac{f(x_i)}{g(x_i)}\right] \tag{4.48}$$

From Eq. (4.48) we see that the variance of the estimator $\overline{p}$ depends on the selection of the importance function. Therefore, this selection is of great importance regarding the convergence speed and the accuracy of the simulation. A wrong selection of the importance function may lead to erroneous results.

4.3.2   *Latin Hypercube Sampling (LHS)*

The method of Latin hypercube sampling (LHS) belongs to the family of stratified sampling techniques. In conventional random sampling new sample points are generated without taking into account the previously generated sample points and without necessarily knowing beforehand how many sample points are needed. In LHS this is not the case, one must first decide how many sample points to use and for each

Fig. 4.5 In LHS method, the samples are randomly generated sampling once from each bin

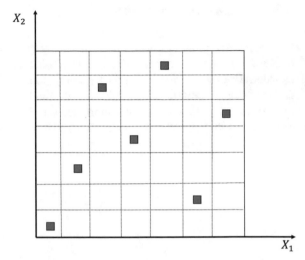

sample point remember its general position. This sampling approach ensures that the samples generated are better distributed in the sample space compared to those obtained by pseudo-random generators. LHS has two simple steps as follows:

1. Divide the range of each random variable into N bins with equal probability mass.
2. Generated one sample from each of the N bins.

Figure 4.5 shows schematically the stratified sampling of two random variables, X_1 and X_2. More specifically, Fig. 4.5 shows how a single-value of X_1 and X_2 is sampled from each of the seven equal-sized bins in order to generate seven LHS samples.

The advantage of this method is that it usually converges much faster than the crude Monte Carlo simulation and reduces the variance of statistical estimates, while the random variables are sampled from the complete range of their possible values, thus ensuring that no subdomain is over-sampled.

4.4 Monte Carlo Methods in Reliability Analysis

4.4.1 Crude Monte Carlo Simulation

If $\mathbf{X} = (X_1, \ldots, X_n)$ is the vector of basic random variables and $G(\mathbf{X})$ is the failure function then for the reliability analysis the calculation of integral of Eq. (4.32) is required. According to MCS an unbiased estimator of the probability of failure is given by the following relation:

$$P_f = \lim_{N \to \infty} \sum_{i=1}^{N} [I(\mathbf{X}_i)] \tag{4.49}$$

where

$$I(\mathbf{X}_i) = \begin{cases} 1 & \text{if } G(\mathbf{X}_i) \leq 0 \\ 0 & \text{otherwise} \end{cases}$$

According to this estimator, N independent random samples from a specific probability distribution are drawn and the failure function is estimated for every one of these samples. If $G(\mathbf{X}_i \leq 0)$ then we have a successful simulation (hit). The probability of failure can be expressed in terms of mean sample value (statistical definition of probability) as follows:

$$P_f = \frac{N_H}{N} \tag{4.50}$$

where N_H is the number of successful simulations and N the total number of simulations.

4.4.2 Importance Sampling

The probability of failure for importance sampling is written, in terms of mean sample value as follows:

$$P_f = \sum_{i=1}^{N} \frac{S_i}{N} \tag{4.51}$$

where

$$S_i = \begin{cases} G(\mathbf{X}_i)/g(\mathbf{X}_i) & \text{if } G(\mathbf{X}_i) \leq 0 \\ 0 & \text{otherwise} \end{cases}$$

and $g(\mathbf{X}_i)$ is the importance function. It is reminded here that importance sampling is based on the idea of concentrating the samples in regions that have a more significant contribution in the final result, balancing the introduced error by reducing the number of successful simulations. This reduction takes place by replacing the unit that is counted for every successful simulation with the weight coefficient $G(\mathbf{X}_i)/g(\mathbf{X}_i)$.

4.4.3 The Subset Simulation (SS)

The estimation of small failure probabilities P_f in high dimensions with the aid of MCS requires an excessive number of simulations in order to capture rare events. The basic idea of subset simulation (SS) is the subdivision of the failure event $G(\mathbf{X}) = \{G(R - S) \leq 0\}$ into a sequence of M partial failure events (subsets)

Fig. 4.6 Division of
$G(\mathbf{X}) = G(X_1, X_2)$ into M
subsequent subsets G_i

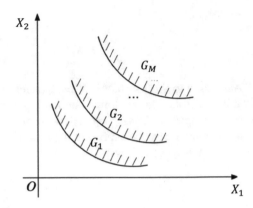

$G_1 \supset G_2 \supset \cdots \supset G_M = G(\mathbf{X})$ of Fig. 4.6. The division into subsets (subproblems) offers the possibility to transform the simulation of rare events into a set of simulations of more frequent events. The determination of the failure events G_i can be expressed as follows:

$$G_i = \{\mathbf{X} : G(\mathbf{X}) \leq b_i\}, \quad i = 1, \ldots, M \tag{4.52}$$

where $b_1 > b_2 > \cdots > b_M$ is a decreasing sequence of threshold values, $\mathbf{X}$ being the vector of independent and identically distributed (i.i.d) samples with probability density function pdf $f_{\mathbf{X}}(\mathbf{X})$.

According to the definition of conditional probability (see Sect. A.3), the computation of the failure probability can be expressed as a product of conditional probabilities $P(G_{i+1}|G_i$ and $P(G_1)$ as follows:

$$P_f = P(G_M) = P(G_1) \cdot \prod_{i=2}^{M} P(G_i|G_{i-1}) \tag{4.53}$$

The determination of the partial failure events G_i and the partial conditional failure probabilities $P_i = P(G_{i+1}|G_i)$ is of great importance for the successful implementation of SS since it strongly affects the accuracy of the simulation. Usually, the limit values $G_i|i = 1, \ldots, M$ are selected in such way that nearly equal partial failure probabilities $P_i|i = 1, \ldots, M$ are obtained for each subset. However, it is difficult to specify in advance the limit-state values G_i according to a prescribed probability P_i and therefore they have to be determined adaptively within the simulation.

Algorithm Implementation of SS

Step 1. Direct MCS with N_1 samples in order to estimate the probability P_1 for subset 1 as follows:

$$P_1 = P(G_1) = \frac{1}{N_1} \cdot \sum_{k=1}^{N_1} I_{G_1}(\mathbf{X}_k^{(1)}) \tag{4.54}$$

where $[\mathbf{X}_k^{(1)} : k = 1, \ldots, N_1]$ are i.i.d samples drawn from $f_{\mathbf{X}}(\mathbf{X})$ and $I_{G_1}(\mathbf{X}_k^{(1)})$ is the indicator

$$I_{F_1}(\mathbf{X}_k^{(1)}) = \begin{cases} 0, & if \ \mathbf{X}_k^{(1)} \notin G_1, \\ 1, & if \ \mathbf{X}_k^{(1)} \in G_1. \end{cases} \tag{4.55}$$

Step 2. Generation of N_i samples of subset i distributed according to

$$\pi(\mathbf{X}|G_i) = \frac{I_{G_{i-1}}(\mathbf{X}) f_{\mathbf{X}}(\mathbf{X})}{P(G_{i-1})} \tag{4.56}$$

in order to obtain the conditional probabilities $P(G_i|G_{i-1})$ of Eq. (4.53). These samples are generated numerically using the Markov Chain Monte Carlo (MCMC) procedure and, in particular, the modified Metropolis-Hastings algorithm is described in Appendix C.

The conditional probability at subset level i can be estimated as

$$P_i = P(G_i|G_{i-1}) = P_i \approx \frac{1}{N_i} \sum_{k=1}^{N_i} I_{G_i}(\mathbf{X}_k^{(i)}) \tag{4.57}$$

where $\mathbf{X}_k^{(i)}$ are i.i.d samples drawn from the conditional pdf $\pi(\mathbf{X}|G_{i-1})$. The N_i samples of subset i are generated from the $P_{i-1} \times N_{i-1}$ samples of subset $i-1$ that are located in the failure region G_{i-1} (seeds) as follows:

$$\textbf{Seeds for subset } i \rightarrow \{\mathbf{X}^{i-1} : G(\mathbf{X}^{i-1}) > G_{i-1}\} \tag{4.58}$$

The number of samples N_i at each subset is selected in a way that the partial probabilities P_i are estimated within an acceptable coefficient of variation (CoV) tolerance, in the context of MCS. The last failure probability $P(G_M|G_{M-1})$ can be estimated with the following expression:

$$P_M = P(G_M|G_{M-1}) \approx \frac{1}{N_M} \sum_{k=2}^{N_M} I_{F_M}(\mathbf{X}_k^{(M)}) \tag{4.59}$$

and the total failure probability P_F may then be computed as

$$P_F = \prod_{i=1}^{M} P_i \tag{4.60}$$

The derivation of the conditional probability estimators of SS (CoV estimates of P_f) is given in Appendix C.

4.5 Artificial Neural Networks (ANN)

In addition to the aforementioned variance reduction methods, surrogate models, such as artificial neural networks (ANN) have been successfully implemented in the framework of reliability analysis leading to cost-efficient yet acceptable predictions of the probability of failure. The associative memory properties featured by these mathematical models allow them to become efficient surrogates to the numerical solver of the mechanical model which is repeatedly invoked in the MCS. ANN is inspired by biological neural networks and are information-processing models configured for a specific application through a training process (see Fig. 4.7).

The first studies describing ANNs were introduced by McCulloch and Pitts (1943, 1947) and Hebb (1949) pioneering work but the idea was soon abandoned. It was not until 1982 that ANN returned to the scene (e.g., Hopfield 1982). The basic property of the ANN is that they adapt efficiently to the input/output relations of complex systems. Their advantage, when applied in the framework of reliability analysis, is that they allow for more accurate mapping of the basic random variables to the response failure surface than the corresponding polynomial approximations (e.g., FORM, SORM), achieving results comparable to brute force MCS with significantly lower computational cost, especially for complex failure domains.

4.5.1 Structure of an Artificial Neuron

ANN models have been widely used for patter recognition, clustering, function approximation, and optimization. In direct analogy to its biological counterpart, an ANN is a cluster of artificial neurons. Every neuron (see Fig. 4.8) is a simple mathematical model consisting of two parts: the net and the activation function. The first determines how the input data are combined in the body of the neuron, while the second determines the output. The net function, multiplies the input data, i.e., x_i, $i = 1, \ldots, n$ by respective weights w_{ij} (similar to synapses in biological neural network), which correspond to the strength of the influence of neuron j, and the resulting values are summed and shifted according to

$$a_j = \sum_{i=1}^{n} x_i w_{ij} + b \qquad (4.61)$$

The constant term b in Eq. (4.61) is the bias value which increases or decreases the summation by a constant so that the neuron may cover a target input range. The initial values of the weights are randomly chosen. In the second part, the activation function $\Phi(\cdot)$ processes the summation result a_j and gives the output y_j of neuron j as follows:

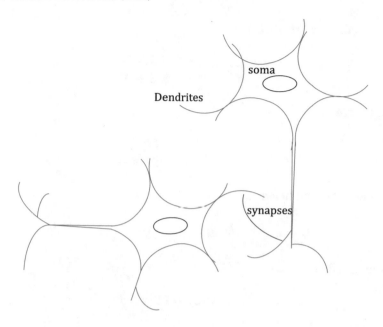

Fig. 4.7 Structure of a biological neuron

$$y_j = \Phi\left(\sum_{i=1}^{n} x_i w_{ij} + b\right) = \Phi(b + \mathbf{w}^\mathsf{T}\mathbf{x}) \tag{4.62}$$

Activation functions are essential parts of the neural network as they introduce the nonlinearity to the network. Some commonly used activation functions are

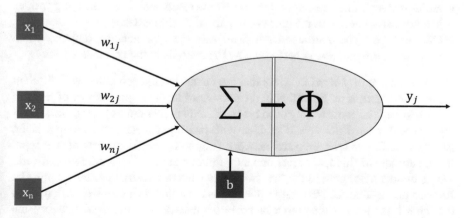

Fig. 4.8 Basic structure of an artificial neuron

- **Step** function (or **threshold**)

$$\Phi(S) = \begin{cases} 1 & \text{if } S > 0 \\ 0 & \text{if } S \leq 0 \end{cases}$$

- **Sign** function

$$\Phi(S) = \begin{cases} 1 & \text{if } S > 0 \\ -1 & \text{if } S \leq 0 \end{cases}$$

- **Sigmoid** or **logistic** function

$$\Phi(S) = \frac{1}{1 + e^{-aS}}$$

- **Linear** function

$$\Phi(S) = \lambda S$$

4.5.2   Architecture of Neural Networks

The architecture of an ANN defines how its several neurons are arranged in relation to each other. In general, an ANN can be divided into three parts as follows:

- **Input** layer: In this layer, the information which is to processed by the neurons of subsequent layers, is provided to the network. In most cases these inputs are normalized within the values of the activation function.
- **Hidden** layers: The neurons of these layers are responsible for identifying various patterns associated with the processing path of the input data.
- **Output** layer: The neurons of this layer give the final network outputs, which result from the processing performed by the neurons in the previous layers.

The neurons are organized in layers that are connected to each other with different patterns and there is no limitation in the number of layers or the number of neurons in each layer. The most widely used types of ANN architectures are: (i) the **single-layer feed-forward** network (Fig. 4.9a, with one input layer and an output layer, (ii) the **multilayer feed-forward** network (Fig. 4.9b, which consists of one input layer, subsequent (hidden) layers and an output layer and (iii) **recurrent** network. The networks belonging to the first two types have one restriction: inputs must be forward propagated, i.e., data and calculations must flow in a single direction, from the input layer toward the output layer. In the networks of the third category, the outputs of the neurons are used as feedback inputs for other neurons.

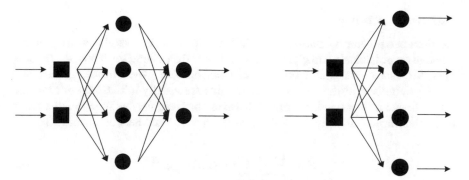

Fig. 4.9 Two simple feed-forward ANNs, **a**—A multilayer network, and **b**—a single-layer network. The *square nodes* represent the input nodes while the *circular nodes* represent the basic processing neurons

4.5.3 Training of Neural Networks

One of the most relevant features of artificial neural networks is their capability to learn the model's behavior by utilizing all available training samples (input/output pairs). This means that after creating a mapping between the input and the output the ANN can produce an output which is close to the expected (or desired) output of any given input values. The ability of the ANN to produce these solutions is called generalization. Therefore, the training of ANN is performed using a learning algorithm which consists of ordinary steps required in order to tune the synaptic weights **w**. Usually, the training data is divided into two subsets. The first subset is the training set and contains samples selected randomly from the total number of samples (around 75–85%) and is used for the training process. The second subset is the test and contains the remaining samples which are used in order to check the networks generalization capability, i.e., generalizing solutions are within acceptable levels.

The three major learning strategies are: (i) **supervised** learning, in which each training sample is composed of the input signals and their corresponding outputs, (ii) **unsupervised** learning in which prior knowledge of the respective outputs is not required, and (iii) **reinforced** learning in which the learning algorithms adjust the weights relying on information acquired through the interaction with the model. This information is used in order to evaluate the learning performance of the ANN. The supervised learning strategy which is widely used for the implementation of ANN in the framework of reliability analysis is presented next.

Supervised Learning

As discussed previously, during the ANN training phase, the weights **w** are properly tuned in order to obtain a mapping that fits closely the training set. Thus, the training of an ANN can be considered as an optimization problem, where the design variables are the weights **w** of the network. All weights are usually initialized with random values drawn from a standard normal distribution. The objective function that needs to be minimized is usually either the mean sum of squared network error (SSE)

$$E = \frac{1}{2} \sum_{l=1}^{n} \sum_{i=1}^{H} (o_{li}(w_{li}) - y_{lh})^2 \qquad (4.63)$$

where o_{li} is the prediction of the i neuron for the l training sample and is apparently a function of w_{li}, while y_{li} is the corresponding exact value. The numerical minimization algorithm used for the training, generates a sequence of weight parameters **w** through an iterative procedure where the update formula can be written as follows:

$$\mathbf{w}^{(t+1)} = \mathbf{w}^{(t)} + \Delta \mathbf{w}^{(t)} \qquad (4.64)$$

The increment of the weight parameter $\Delta \mathbf{w}^{(t)}$ is further decomposed into

$$\Delta \mathbf{w}^{(t)} = a_t d^{(t)} \qquad (4.65)$$

where d_t is a search direction vector and a_t is the step size along this direction. The algorithms most frequently used in the ANN training are the steepest descent, the conjugate gradient, and the Newton's methods with the following direction vectors:

- Steepest descent method: $d^{(t)} = -\nabla E(\mathbf{w}^{(t)})$
- Conjugate gradient method: $d^{(t)} = -\nabla E(\mathbf{w}^{(t)}) + \beta_{t-1} d^{(t-1)}$ with $\beta_{t-1} = (\nabla E_t)^2 / (\nabla E_{t-1})^2$
- Newton's method: $d^{(t)} = -[E(\mathbf{w}^{(t)})]^{-1} \nabla E(\mathbf{w}^{(t)})$

A very famous learning algorithm based on an adaptive version of the Manhattan-learning rule is the **R**esilient back**prop**agation abbreviated as **Rprop**. In Rprop, the ANN learning process progresses iteratively, through a number of epochs (which is defined as a complete processing of all the samples of the training set). On each epoch the error of Eq. (4.63) is calculated by comparing the actual outputs with the corresponding target values. The basic idea of Rprop is that if the partial derivative $\partial E / \partial w$ is negative, the weight is increased (left part of Fig. 4.10) and if the partial derivative is positive, the weight is decreased (right part of Fig. 4.10). This ensures that a local minimum is reached.

The weight updates can be written

$$\Delta w_{ij}^{(t+1)} = -\eta_{ij}^{(t+1)} \text{sgn} \left(\frac{\partial E_t}{\partial w_{ij}} \right), \qquad (4.66)$$

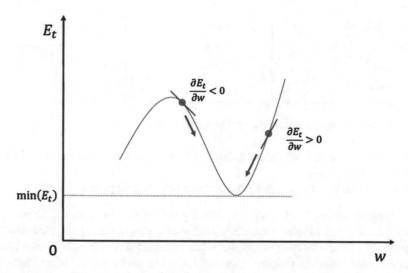

Fig. 4.10 Basic idea of the backpropagation algorithm for a univariate error function $E_t(w)$

where

$$
\eta_{ij} = \begin{cases} \min(\alpha\eta_{ij}^{(t)}, \eta_{\max}), & \text{if } \frac{\partial E_{t+1}}{\partial w_{ij}} \frac{\partial E_t}{\partial w_{ij}} > 0 \\ \max(b\eta_{ij}^{(t)}, \eta_{\min}), & \text{if } \frac{\partial E_{t+1}}{\partial w_{ij}} \frac{\partial E_t}{\partial w_{ij}} < 0 \\ \eta_{ij}^{(t-1)} & \text{otherwise} \end{cases} \tag{4.67}
$$

where $a = 1.2, b = 0.5$, and η are step parameters called "locally adaptive" learning rates because they are based on weight specific information, such as the temporal behavior of the partial derivative of this weight. The learning rates are bounded by upper and lower limits in order to avoid oscillations and arithmetic underflow.

During the ANN training, large weights can cause an excessively large variance of the output. A way of dealing with the negative effect of large weights is regularization. The idea of regularization is to make the network response smoother through modifying the objective function by adding a penalty term that consists of the squares of all network weights. This additional term favors small values of weights and minimizes the tendency of the model to overfit.

The Problem of Overfitting

One of the problems that occur during neural network training is the over-fitting (see Fig. 4.11), which occurs when the error on the training set is driven to a very small value, but when new data is presented to the network the error is large. In order to prevent overfitting, we must improve network's generalization. This can be achieved by the following:

1. Stop the training early - before it "learns" the training data too well.
2. Retrain several neural networks

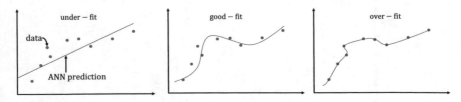

Fig. 4.11 Training performance of an ANN over a set of data

3. Add some form of regularization term to the error function to encourage smoother network mappings.
4. Add noise to the training patterns to smear out the data points.

The computational cost of training an ANN depends on the dimension of the problem. More specifically, the higher the dimensionality of the parameter space, the higher the time required for the training because of a large number of mathematical calculations required to estimate the weights. However, in cases where the numerical model is computationally demanding (for example a nonlinear dynamic model of a large-scale structure), the computational cost of training an ANN is only a small fraction of the total computational effort.

4.5.4 ANN in the Framework of Reliability Analysis

In the context of reliability analysis, ANN models are used as surrogates to the detailed FEM models. This is accomplished by considering the vector $\mathbf{X}$ of n basic random variables as input to the ANN and the corresponding values of the limit-state function $G(\mathbf{X})$ as output. More specifically, the basic steps for implementing the ANN technique for the estimation of the failure probability P_f of a FEM model are as follows:

Step 1. Given the probabilistic characteristics of the random variables $\mathbf{X}$ generate N samples from the joint pdf of $\mathbf{X}$ together with its corresponding pair $G(\mathbf{X})$, i.e.,

$$\mathbf{X}_1 = [X_1^{(1)}, X_2^{(1)}, \ldots, X_n^{(1)}], \quad G(\mathbf{X}_1)$$
$$\mathbf{X}_2 = [X_1^{(2)}, X_2^{(2)}, \ldots, X_n^{(2)}], \quad G(\mathbf{X}_2)$$
$$\vdots$$
$$\mathbf{X}_N = [X_1^{(N)}, X_2^{(N)}, \ldots, X_n^{(N)}], \quad G(\mathbf{X}_N)$$

These tuples $\{\mathbf{X}_i, G(\mathbf{X}_i)\}$, $i = 1, \ldots, N$ are the data used for the ANN training and validation.

Step 2. Using $N_{\text{train}} = (75\text{–}80\%)N$ randomly selected samples, train the ANN using the methods described in Sect. 4.5.3.

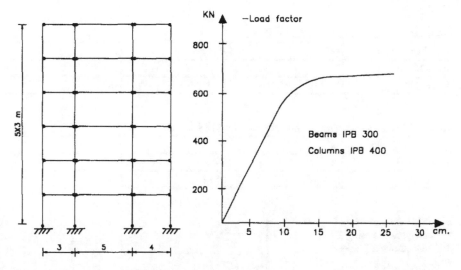

Fig. 4.12 Five-storey plane frame with data; loading, mode of failure, and load-displacement curve

Step 3. Use the remaining $N_{test} = N - N_{train}$ samples in order to check the performance of the trained ANN in generalization. In order to do that, estimate the mean squared error

$$e^{MSE} = \frac{1}{N_{test}} \sum_{j=1}^{N_{test}} ||G(\mathbf{X}_j) - \hat{G}(\mathbf{X}_j)||_2 \tag{4.68}$$

where $G(\cdot)$ and $\hat{G}(\cdot)$ are the exact and approximated values of the limit-state function for any realization of $\mathbf{X}$. If e is less than a threshold then use the ANN to estimate P_f, else repeat step 2 using a different subset of the available data.

Step 4. Generate a large number N_{MCS} of realizations of $\mathbf{X}$ and calculate the corresponding limit-state functions using the ANN instead of the detailed FEM model. The probability of failure is then estimated according to Eq. (4.50) as follows:

$$P_f = \frac{N_{\hat{H}}}{N_{MCS}} \tag{4.69}$$

where $N_{\hat{H}}$ is the number of samples for which $\hat{G}(\mathbf{X}) \leq 0$.

The use of ANN in the framework of reliability analysis of large-scale models can be prohibitive due to the large dimensionality of the required input vector in the ANN training process. For example, in the case of an FEM model with 10^6 finite elements which utilizes the midpoint method for the random field mapping, an input vector of up to 10^6 random variables would be required in the ANN architecture. This number is impossible to be handled by any ANN.

Table 4.1 Characteristics of random variables

Random variable	pdf	μ	σ
σ_y (kN/cm)	N	24	2.4
Load (kN)	Log-N	6.57	0.2

Table 4.2 Characteristics of random variables for MCS-IS

Random variable	pdf	μ	σ
σ_y (kN/cm)	N	24	2.4
Load (kN)	Log-N	855	165

Fig. 4.13 Performance of NN configuration using different number of hidden units

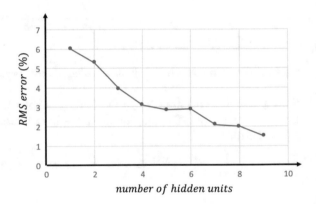

4.6   Numerical Examples

1. A horizontal unit load is applied at the top storey of the frame of the five-storey plane frame shown in Fig. 4.12. The dimensions and section properties of the frame, the load-displacement curve until the formation of the collapse mechanism, as well as the plastic nodes at collapse, using the mean values of the basic random variables are also depicted in Fig. 4.12.

   The yield stress and the external load are considered to be random variables with specified probability density function depicted in Tables 4.1 and 4.2 for brute MCS and MCS-IS, respectively. The loads acting on the structure follow a Log-normal probability density function, while the yield stress follow a normal probability density function.

   We want to test the performance of ANN in estimating the probability of failure which is represented by the total collapse of the structure due to the successive formation of plastic nodes. The "exact" probability of failure has been computed with conventional MCS. Twenty values of the yield stress were used as input variables for the limit elastoplastic analysis. Ten of them were selected, together with their corresponding "exact" critical load factors produced by the limit elastoplastic analysis, to be the training set. The remaining "exact" pairs are used to test the accuracy of the NN prediction.

Figure 4.13 demonstrates the performance of the ANN configuration using different numbers of hidden units. It can be seen that the RMS error is reaching a plateau after a certain number of hidden units without any further improvement. After the selection of the best ANN architecture, the network was tested for different generalization tolerances, in order to examine the existence of any overtraining effects before choosing the final trained ANN. Once an acceptable trained ANN in predicting the critical load factors is obtained, the probability of failure for each test case is estimated by means of ANN-based Monte Carlo Simulation using the Basic MCS and the MCS with IS. The results and performance of ANN compared to corresponding MCS are depicted in Tables 4.3 and 4.4.

From these tables it can be observed that, in the case of basic MCS simulation, the maximum difference of the predicted probability of failure with respect to the "exact" one is 30%, while the corresponding difference in the case of MCS with IS is around 10%, while the computational effect required for the ANN approach is two orders of magnitude less.

Table 4.3 "Exact" and predicted values of P_f

Number of simulations	"exact" MCS	ANN MCS	"exact" MCS-IS	ANN MCS-IS
	P_f	P_f	P_f	P_f
50	6.00	2.00	8.18	6.72
100	7.00	5.00	7.45	6.28
300	8.00	5.33	7.88	6.72
500	8.20	5.40	8.61	7.30
1000	8.60	5.90	8.55	7.38
5000	8.48	5.98	8.46	7.40
10000	8.36	5.93	8.44	7.40
50000		6.04		7.38
100000		6.04		7.38

Table 4.4 Required CPU time for the prediction of P_f. (⋆ corresponds to 10000 MCS)

CPU time in seconds	"exact MCS"	ANN-MCS	"exact"-MCS	ANN-IS
Pattern recognition	–	6	–	6
Training	–	4	–	6
Propagation	–	20	–	20
Total	3134⋆	30	3134⋆	30

4.7   Exercises

1. Given the limit-state function $g(X_1, X_2) = X_1 + 2X_2 - 24$ where $X_1 \sim N(0, 4)$ and $X_2 \sim N(3, 1)$ are correlated Gaussian $N(\mu, \sigma^2)$ random variables with covariance matrix

$$\mathbf{C} = \begin{bmatrix} 0.4 & 0.3 \\ 0.3 & 0.9 \end{bmatrix}$$

   calculate reliability index β.

2. For the problem of previous exercise repeat the calculations for the case of the nonlinear failure function $g(X_1, X_2) = X_2^2 + 2X_2 + 24(X_1 + X_2)$. (**Hint**: use a newton–raphson method to converge to the solution of the gradient-based optimization problem of Eqs. (4.26) and (4.27).)

3. For the cantilever of Fig. 3.10 compute the probability of $u_{max} \geq 0.4$ m using brute force MCS.

4. Repeat the calculations of example 3 using subset simulation and compare the results. (**Hint**: Compute intermediate failure events with lower probability of $P_i = 0.1$ and generate conditional samples at each subset level using the Metropolis–Hastings algorithm of Appendix C).

Appendix A
Probability Theory

Reasoning: *The process of thinking about something in a logical way in order to form a conclusion or judgement.*

Deductive reasoning or deduction, starts out with a general statement, or hypothesis, or axiom and reaches a specific conclusion with the use of intermediate logical steps. Scientific approaches use deduction to test hypotheses and theories as well as to create mathematically sound predictive models of physical phenomena. In deductive reasoning, if something is true of a class of things in general, it is true for all members of that class. Thus, deductive reasoning provides with absolute proof of the conclusions, provided that the premises (axioms) are true. For example, **Axiom**: All lemons are yellow. **Conclusion**: A fruit that is blue is NOT a lemon.

Inductive reasoning is the opposite of deductive reasoning. Inductive reasoning makes broad generalizations from specific observations (data). In other words, inductive reasoning postulates a general conclusion or a theory that could explain the data and by no means it provides a proof that this theory is correct. Even if all of the premises are true in a statement, inductive reasoning allows for the conclusion to be false. Inductive reasoning has its place in scientific approaches. Scientists use it to form hypotheses and theories, which are then fed to deductive reasoning for further processing. For example, the observed probability of lung cancer is very high among smokers. Therefore, a reasonable, but yet not certain, conclusion is that smoking causes lung cancer.

The origin of probability theory lies in physical observations associated with games of chance and was initially based on induction. In many cases of repeated trials, it was observed that after a large number of trials, some quantities remained constant. It is as late as the beginning of the twentieth century that a connection of probability theory to mathematical set theory was established, leading to the foundation of a deductive theory by means of the axiomatic definition of probability.

© Springer International Publishing AG 2018
V. Papadopoulos and D.G. Giovanis, *Stochastic Finite Element Methods*,
Mathematical Engineering, https://doi.org/10.1007/978-3-319-64528-5

A.1   Axiomatic Probability Theory

Axiomatic probability theory is based on the mathematical **set theory** of well-determined object collections, called sets. Set theory, as a mathematical discipline, begins in the work of **Georg Cantor** in 1870s. The objects of a set are called **elements**. The basic terminology used in set theory is presented next:

- **Sample space** (Θ): Set of all possible outcomes of an experiment
  1. Throwing a die (Fig. A.1): $\Theta = \{1, 2, 3, 4, 5, 6\}$
  2. Throwing two dice: $\Theta = \{\{1, 1\}, \{2, 6\}, \ldots \{5, 6\}, \{6, 6\}\}$

- **Events**: Subsets of the sample space

- **Sample point** (θ): Elements of the sample space Θ

- **Null (empty) set** ($\emptyset$): Set that contains no elements.

Two sets A and B are identical **if and only if** they have they same elements. The number of the elements of set A is called **cardinality** of the set and is denoted as $|A|$ or $card(A)$. The set of all subsets of A is called **power set** and is denoted as $\mathfrak{P}(A)$ or 2^A.

Example: if $A = \{a, b\}$ then $\mathfrak{P}(A) = \{\emptyset, \{a\}, \{b\}, \{a, b\}\}$.

A.1.1   Basic Set Operations

For two sets A and B, the following axiomatic definitions are given:

Set inclusion($\subset$): A set A is included in set B or it is a subset of B if all elements of A are also elements of B.

$$(A \subset B) \tag{A.1}$$

Union($\cup$): The union of sets A and B is the set $A \cup B$ of all elements that belong to A or B, or both. It is denoted by $A \cup B$ and expressed as

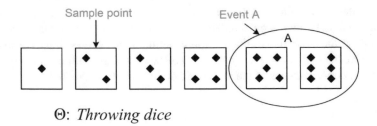

Θ: *Throwing dice*

Fig. A.1 The sample space of throwing dice

$$A \cup B = \{\omega | \omega \in A \ \ or \ \ \omega \in B\} \tag{A.2}$$

Intersection($\cap$): The intersection of sets A and B is the set $A \cap B$ of all elements that belong to both A and B. It is denoted by $A \cap B$ and expressed as

$$A \cap B = \{\omega | \omega \in A \ \ and \ \ \omega \in B\} \tag{A.3}$$

Mutually exclusive: The sets A and B are called mutually exclusive if their intersection is the null set

$$A \cap B = \emptyset \tag{A.4}$$

Complement: The complement of a set A is the set with all the elements that are not in A

$$A^c = \{\omega | \omega \notin A\} \tag{A.5}$$

Difference: The difference of sets A and B is the set of all elements that belong to A but that do not belong to B

$$A - B = A \cap B^c \tag{A.6}$$

Complement of the union: The complement of the union of two sets is the intersection of their complements

$$(A \cup B)^c = A^c \cap B^c \tag{A.7}$$

A.1.2 Set Equality Theorems

There are a number of general theorems about sets which follow from the definitions of set-theoretic axioms. The equations below hold for any sets X, Y, Z:

1. **Idempotent laws**:
   (a) $X \cup X = X$
   (b) $X \cap X = X$
2. **Commutative Laws**:
   (a) $X \cup Y = Y \cup X$
   (b) $X \cap Y = Y \cap X$
3. **Associative Laws**:
   (a) $(X \cup Y) \cup Z = X \cup (Y \cup Z)$
   (b) $(X \cap Y) \cap Z = X \cap (Y \cap Z)$
4. **Distributive Laws**:
   (a) $X \cup (Y \cap Z) = (X \cup Y) \cap (X \cup Z)$
   (b) $X \cap (Y \cup Z) = (X \cap Y) \cup (X \cap Z)$

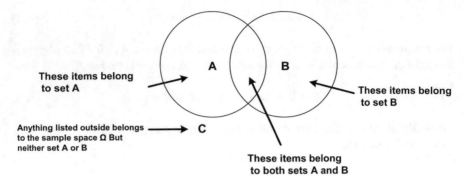

Fig. A.2 A Venn diagram is a tool used to show intersections, unions, and complements of sets

5. **Identity Laws**:
   (a) $X \cup \emptyset = X$ (b) $X \cap \emptyset = \emptyset$
   (a) $X \cup \Omega = \Omega$ (b) $X \cap \Omega = X$
6. **Complement Laws**:
   (a) $X \cup X^c = \Omega$ (b) $X \cap X^c = \emptyset$
   (a) $(X^c)^c = X$ (b) $X - Y = X \cap Y^c$
7. **De Morgans Laws**:
   (a) $(X \cup Y)^c = X^c \cap Y^c$
   (b) $(X \cap Y)^c = X^c \cup Y^c$.

A.1.3   Venn Diagrams

Events and sets can be represented as spaces that are bounded by closed shapes called **Venn diagrams** and are used in order to graphically represent all possible logical relations between a finite collection of different sets. Venn diagrams were conceived by **John Venn** (1880). A Venn diagram is depicted in Fig. A.2.

A.2   Definitions of Probability

A.2.1   Classical Definition

The classical definition or interpretation of probability is identified with the works of **Jacob Bernoulli** and **Pierre–Simon Laplace**. As stated in Laplace's *Théorie analytique des probabilités* (1886),

The probability of an event is the ratio of the number of cases favorable to it, to the number of all cases possible when nothing leads us to expect that any one of these cases should occur more than any other, which renders them, for us, equally possible.

This definition is essentially a consequence of the **principle of indifference** which states that if elementary events are assigned equal probabilities, then the probability of a disjunction of elementary events is just the number of events in the disjunction divided by the total number of elementary events. The probability of an event A written as $P(A)$ can be written as a fraction where the numerator is the number of favorable outcomes $N(A)$ and the denumerator is the number of possible cases $N(\Theta)$

$$P(A) = \frac{N(A)}{N(\Theta)} \tag{A.8}$$

Drawbacks

- All possible outcomes are considered equal-probable
- This definition stands for random experiments with finite number of possible outcomes.

A.2.2 Geometric Definition

Geometric probability is an extension of the classical definition in experiments with infinite number of possible outcomes. Some problems, like **Buffon's needle**,[1] **Birds On a Wire**, **Bertrand's Paradox**, or the problem of the **Stick Broken Into Three Pieces** do, by their nature, arise in a geometric setting. Geometric probabilities are nonnegative quantities not exceeding the value 1 which are assigned to subregions of a given domain, and are subjected to certain rules. The probability of selecting inside a subregion A of a domain Ω in $\mathbb{R}^n$ can be defined as

$$P(A) = \frac{\mu(A)}{\mu(\Theta)} \tag{A.9}$$

where $\mu(\cdot)$ can be length, area, volume, etc., of the corresponding subregion.

---

[1] Is one of the oldest problems in the field of geometrical probability (see Fig. A.3). It was first posed in the 18th century by **Georges-Louis Leclerc, Comte de Buffon**: It involves dropping a needle on a lined sheet of paper and determining the probability that the needle crosses one of the lines on the page (Fig. A.3). The result is that the probability is directly related to the value of π (Schroeder 1974).

Fig. A.3 Buffon's needle
illustration

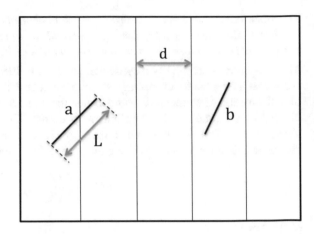

A.2.3   Frequentist Definition

In an effort to go beyond the limitations posed by the condition of equalprobable
of the possible outcomes of an experiment standing in the classical definition of
probability fields statistical probability was presented in the 1920s by **Richard Von
Mises** (1939) but the frequentist view may have been foreshadowed by Aristotle, in
Rhetoric, when he wrote:

"The probable is that which for the most part happens"

Frequentist probability defines an event's probability as the limit of its relative frequency in a large number of trials:

$$P(A) = \lim_{n \to \infty} f_n(A) = \lim_{n \to \infty} \frac{\mu(A)}{n} \tag{A.10}$$

In this **frequentist** interpretation, probabilities are defined over experiments that are
conducted under the **exact same conditions**, which is practically impossible.

A.2.4   Probability Space

A **probability space** is a mathematical triplet $(\Theta, \mathscr{F}, P)$ containing the **sample
space** Θ, the σ-**algebra** $\mathscr{F}$ and the **probability measure** P. In mathematical analysis
and probability theory, a σ-algebra $\mathscr{F}$ on a set A is a collection of subsets of A that
is closed under countable fold set operations. Let A be some set, and let 2^A represent
its power set. Then a subset $\mathscr{F} \subseteq 2^A$ is called a σ-algebra if it satisfies the following
three properties:

1. $\mathscr{F}$ is non-empty: There is at least one $(X \subset A) \in \mathscr{F}$.
2. $\mathscr{F}$ is closed under complementation: If X is in $\mathscr{F}$, then so is its complement, $X^x \in \mathscr{F}$.
3. $\mathscr{F}$ is closed under countable unions: If $X_1, X_2, X_3, \ldots$ are in $\mathscr{F}$, then so is $X = X_1 \cup X_2 \cup X_3 \cup \ldots$.

The pair $(A, \mathscr{F})$ without P is a field of sets, called a **measurable space**. The probability measure P is a function returning an event's probability. A probability is a real number between zero and one. Thus, P is a function $P : \mathscr{F} \to [0, 1]$.

A.2.5 The Axiomatic Approach to Probability

In the axiomatic definition, the probability P of some event A, denoted $P(A)$, is usually defined such that P satisfies the **Kolmogorov**[2] axioms:

1. The probability of an event A is a positive number $P(A)$ assigned to this event: $P(A) \geq 0$
2. The probability of the certain event Ω equals 1: $P(\Theta) = 1$.
3. If the events $A_1, A_2, \ldots$ are mutually exclusive (there is no element common to both events: $A_1 \cap A_2 \cap \ldots = \emptyset$), then

$$P\left(\bigcup_{i=1}^{\infty} A_i\right) = \sum_{i=1}^{\infty} P(A_i) \tag{A.11}$$

These axioms can be summarized as: Let $(\Theta, \mathscr{F}, P)$ be a measure space with $P(\Theta) = 1$. Then $(\Theta, \mathscr{F}, P)$ is a probability space, with sample space Θ, σ-algebra $\mathscr{F}$ and probability measure P.

Elementary Properties of the Probability

A probability space satisfies the following properties:

- The probability of the null set $\emptyset$ is zero: $P(\emptyset) = 0$
- The probability of an event A is: $0 \leq P(A) \leq 1$
- The probability of a Θ is: $P(\Theta) = 1$
- $P(A^c) = 1 - P(A)$
- If $A \subseteq B$ then $P(A) \leq P(B)$
- If A and B are mutually exclusive: $P(A \cup B) = P(A) + P(B)$. In general: $P(A \cup B) \leq P(A) + P(B)$

The events $A_i, Aj, \ldots, A_n$ are said to be **mutually independent if and only if** the following relation holds

$$P(A_1 \cap A_2 \cap \ldots \cap A_n) = P(A_1)P(A_2) \ldots P(A_n) \tag{A.12}$$

---

[2] Andrey Nikolaevich Kolmogorov (1903–1987).

A.3   Conditional Probability

Let A and B be events in Θ, and suppose that $P(B) > 0$. The probability of the event A given the occurrence of the event B is estimated by considering the conditional probability of A given that B occurs. The **conditional probability** of A given B is by definition the ratio:

$$P(A \mid B) = \frac{P(A \cap B)}{P(B)} \tag{A.13}$$

The following properties follow from the definition:

- If $B \subset A$ then $P(A \mid B) = 1$
- If $A \subset B$ then $P(A \mid B) = \frac{P(A)}{P(B)} \geq P(A)$.

A.3.1   Multiplication Rule of Probability

The rule of multiplication applies in cases were we are interested for the probability of a sequence of events $A_1, A_2, \ldots, A_n$. This probability is equal to the probability that the event A_1 occurs times the probability the event A_2 occurs given that A_1 has occurred times the probability that A_3 occurs given that $A_1 \cap A_2$ has occurred, etc.

$$P(A_1 \cap A_2 \cap \ldots \cap A_n) = P(A_1) \cdot P(A_2|A_1) \cdot \ldots \cdot P(A_n|A_1 \cap A_2 \cap \ldots \cap A_{n-1}) \tag{A.14}$$

A.3.2   Law of Total Probability

In probability theory, the **law of total probability** is a rule that relates marginal probabilities to conditional probabilities. It expresses the total probability of an outcome which is realized via several distinct events. Suppose that $(A_1, A_2, \ldots, A_n)$ is a countable collection of disjoint events that partition the sample space Θ and $P(A_i) > 0$, $i = 1, \ldots, n$. For the event B we get(see Fig. A.4).

$$
\begin{aligned}
B &= (B \cap A_1) \cup (B \cap A_2) \ldots (B \cap A_n) \rightarrow \text{disjoint} \\
P(B) &= P(B \cap A_1) + P(B \cap A_2) + \ldots P(B \cap A_n) \rightarrow \text{multipl.} \quad \text{(A.15)} \\
P(B) &= P(B|A_1)P(A_1) + P(B|A_2)P(A_2) + \ldots P(B|A_n)P(A_n) \\
&= \sum_{i=1}^{n} P(A_i)P(B|A_i)
\end{aligned}
$$

Fig. A.4 Partition of a
sample space Ω in a
countable collection of
events

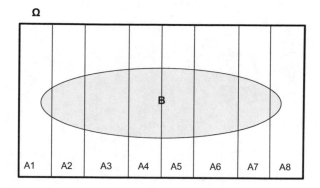

A.3.3   Bayes Theorem

Bayes theorem can be seen as a way of understanding how the probability is affected
by new pieces of evidence. The theorem relates the conditional probability $P(A_i \mid B)$
and the probabilities $P(A_i)$, $P(B)$ of events A and B, provided that the probability
of B does not equal zero:

$$P(A_i|B) = \frac{P(A_i)P(B|A_i)}{\sum_{i=1}^{n} P(A_i)P(B|A_i)} = \frac{P(A_i)P(B|A_i)}{P(B)} \qquad (A.16)$$

In Bayes theorem, each probability has a conventional name: $P(A_i)$ is the prior
probability (or "unconditional" or "marginal" probability) of A_i in the sense that
it does not take into account any information about B, $P(A_i|B)$ is the conditional
(posteriori) probability of A_i, given B, because it is derived from or depends upon
the specified value of B, $P(B|A_i)$ is the conditional probability of B, given A_i and
$P(B)$ is the prior or marginal probability of B.

Appendix B
Random Variables

A random variable $X(\theta)$, $\theta \in \Theta$ is a function defined on a sample space Θ, i.e., the space of all possible outcomes θ of a random experiment, such that for every real number x the probability $P[\theta : X(\theta) \leq x]$ exists. In more simple terms a random variable is a rule that assigns a real number to each possible outcome of a probabilistic experiment (Fig. B.1). For example, random variable can be considered the height of a randomly selected student in a class or the value of the Young's modulus of a structure.

Random variables can be either discrete, that is, taking any of a specified finite or countable list of values, and/or continuous, that is, taking any numerical value in an interval or collection of intervals.

Fig. B.1 Definition of a random variable as a mapping from $\Theta \rightarrow \mathbb{R}$

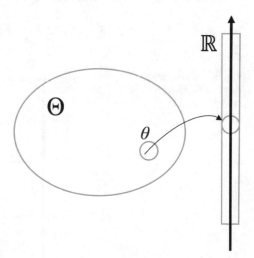

V. Papadopoulos and D.G. Giovanis, *Stochastic Finite Element Methods*,
Mathematical Engineering, https://doi.org/10.1007/978-3-319-64528-5

B.1   Distribution Functions of Random Variables

B.1.1   Cumulative Distribution Function (cdf)

The probability distribution (or cumulative distribution) function (cdf) $F_X(x)$ describes the probability that a random variable X takes on a value less than or equal to a number x (Fig. B.2).

$$F_X(x) = P[X \leq x]$$
$$= P\{\theta \in \Theta, -\infty \leq X(\theta) \leq x\} \tag{B.1}$$

The cdf of a random variable has the following properties:

- $F_X(x)$ is non-decreasing, i.e., if $a < b$ then $F_X(a) \leq F_X(b)$
- The limit of the cdf tends to zero as the value x tends to minus infinity and to the unit as x tends to infinity

$$\lim_{x \to -\infty} F_X(x) = 0, \quad \lim_{x \to +\infty} F_X(x) = 1$$

- $F_X(x)$ is right-continuous, i.e., for a converging sequence x_n such that $x_n \to x^+$,

$$\lim_{n \to \infty} F_X(x_n) = F_X(x)$$

- The probability that x lies between a, b is defined as

$$P[a \leq x \leq b] = F_X(b) - F_X(a). \tag{B.2}$$

Fig. B.2 Cumulative distribution function of a random variable

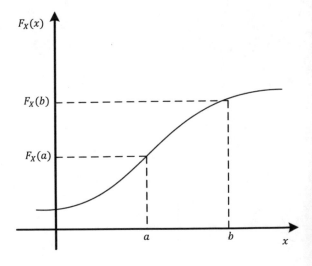

B.1.2 Probability Density Function (pdf)

The probability density function $f_X(x)$ of a random variable X is defined to be the function which when **integrated** yields the **cdf** (Fig. B.3):

$$F_X(x) = \int_{-\infty}^{x} f_X(x)dx \Rightarrow f_X(x) = \frac{dF_X(x)}{dx} \tag{B.3}$$

The probability density function (pdf) can take values greater than one, while if we calculate the area beneath the pdf the value we take is equal to one: $\int_{-\infty}^{\infty} f_X(\xi)d\xi = 1$. A pdf can be unimodal or multimodal (Fig. B.4). The mode is the value that appears most often in a set of data. The probability that the random variable falls within the range $[a, b]$ is estimated as the integral of the random variable's pdf over that range

$$P(a < X \leq b) = F(b) - F(a) = \int_{a}^{b} f_X(x)dx \tag{B.4}$$

This probability is equal to area under the density function but above the horizontal axis and between the lowest and greatest values of the range (Fig. B.3).

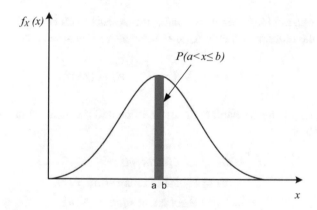

Fig. B.3 The probability $P(a < x \leq b)$ is estimated as the shaded area of the pdf between a and b

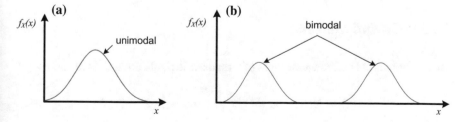

Fig. B.4 **a** A unimodal and **b** a multimodal (bimodal) distribution function

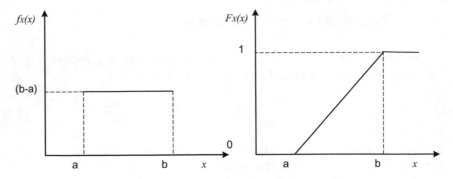

Fig. B.5 The pdf and cdf functions of a random variable with uniform distribution

Figure B.5 depicts the pdf and the corresponding cdf of a random variable uniformly distributed in the range [a, b]. As can be seen in this figure, the pdf is constant, indicating that all values of the random variable in this range are equiprobable, and the resulting cdf is a straight monotonically increasing transition from 0 to 1.

B.2 Moments of Random Variables

As moments of a random variable we define the expected values of powers or related functions of the random variable. An n-th moment can be defined as:

$$m_n(x) = \mathbb{E}[X^n] = \int_{-\infty}^{+\infty} x^n f_X(x)\mathrm{d}x \tag{B.5}$$

The first two moments are namely the mean value and the mean square value of the random variable:

$$\mu_X = m_1 = \mathbb{E}[X] = \textbf{mean}$$
$$m_2 = \mathbb{E}[X^2] = \textbf{mean square}$$
$$\sqrt{m_2} = \textbf{Root mean square (RMS)}$$

$$\tag{B.6}$$

B.2.1 Central Moments

Central moment is defined a moment of a random variable around its mean:

$$K_n(x) = \mathbb{E}[(X - \mu_X)^n] = \int_{-\infty}^{+\infty} (x - \mu_X)^n f_X(x)\mathrm{d}x \tag{B.7}$$

The first central moment which is zero must not be confused with the first moment itself, the expected value or mean, while the second central moment is called the variance, which is the square of the standard deviation:

$$\sigma_X^2 = K_2 = \mathbb{E}[(X - \mu_X)^2] \rightarrow \textbf{variance}$$

$$\sigma_X = \sqrt{\text{variance}} \rightarrow \textbf{standard deviation} \qquad (B.8)$$

It can be easily demonstrated that the variance can be obtained as

$$\textbf{variance} = \sigma_X^2 = \mathbb{E}[(X^2)] - \mu_X^2 \qquad (B.9)$$

Coefficient of Variation (CoV)

Coefficient of variation (CoV) is a measure of the dispersion around the mean value given as

$$\text{CoV} = \frac{\sigma_X}{\mu_X} \qquad (B.10)$$

Coefficient of Skewness

Skewness is a measure of the asymmetry of the probability distribution of a random variable about its mean (Fig. B.6) defined as

$$\gamma_1 = \frac{K_3}{\sigma_X^3} \qquad (B.11)$$

Coefficient of Kurtosis

The coefficient of kurtosis is defined as the fraction

$$\gamma_2 = \frac{K_4}{\sigma_X^4} \qquad (B.12)$$

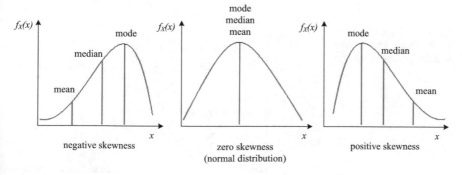

Fig. B.6 Graphs of distributions with negative, zero and positive skewness

Its values are often compared to a **standard value** of 3. The value of 3 corresponds to the value of the **kurtosis coefficient of the Gaussian distribution**. A $(\gamma_2 - 3) > 0$ implies a slim sharp peak in the neighborhood of a mode in a unimodal distribution (sharper than the Gaussian distribution having the same σ_X), while a negative $(\gamma_2 - 3) < 0$ implies as a rule, a flattened peak (flatter than the Gaussian distribution having the same σ_X) (Fig. B.7).

B.3 Functions of Random Variables

Consider the case of two random variables X and Y related as $Y = g(X)$. The question that arises is: "Which is the cumulative distribution of Y in terms of the cumulative distribution of X?"

B.3.1 One to One (1–1) Mappings

The easiest case for transformations of continuous random variables is the case of g to be one to one (1–1) mapping. Consider first the case of g being an increasing function in the range of the random variable X (Fig. B.8), then, g^{-1} is also going to be increasing function. Therefore:

$$F_Y(y) = P\{Y \leq y\} = P\{g(X) \leq y\} = P\{X \leq g^{-1}(y)\} = F_X(g^{-1}(y)) \quad \text{(B.13)}$$

In the case of decreasing g in the range of the random variable X we have

$$F_Y(y) = P\{Y \leq y\} = P\{g(X) \leq y\} = P\{X \geq g^{-1}(y)\} = 1 - F_X(g^{-1}(y))$$
$$\text{(B.14)}$$

Since the events in dx and dy are equiprobable (i.e areas A1 and A2 in Fig. B.8 are equal), it can be easily shown that the probability density function is obtained as

Fig. B.7 Graph of distributions with negative, zero and positive kurtosis

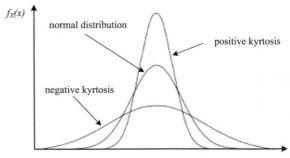

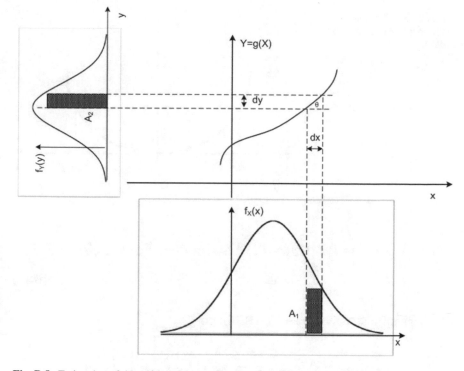

Fig. B.8 Estimation of the pdf for an increasing function of a random variable

$$f_Y(y) = f_X(g^{-1}(y)) \left| \frac{d}{dy} (g^{-1}(y)) \right|. \tag{B.15}$$

B.3.2   Not One to One (1–1) Mappings

In the general case where g is not an one to one (1–1) mapping we can use the **Leibnitz Rule**:

$$\frac{d}{dy} \int_{a(y)}^{b(y)} f(x, y) dx = \int_{a(y)}^{b(y)} \frac{\partial}{\partial y} [f(x, y)] dx + f(b, y) \frac{db}{dy} - f(a, y) \frac{da}{dy} \tag{B.16}$$

Example: Consider two random variables X and Y related as $Y = G(X) = X^2$ (Fig. B.9). Assuming that $f_X(x)$ is given, then, $f_Y(y)$ can be calculated from

Fig. B.9 Graph of the
function of the two random
variables

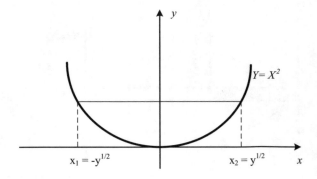

$$F_Y(y) = P[Y \le y] = P[X^2 \le y] = P[-\sqrt{y} \le X \le \sqrt{y}]$$
$$= \int_{-\sqrt{y}}^{\sqrt{y}} f_X(x)\mathrm{d}x \tag{B.17}$$

after implementing Leibniz rule (Eq. (B.16)) as:

$$f_Y(y) = \frac{\mathrm{d}F_Y(y)}{\mathrm{d}y} = f_x(\sqrt{y})\frac{1}{2}y^{-\frac{1}{2}} - f_x(-\sqrt{y})(-\frac{1}{2})y^{-\frac{1}{2}}$$
$$= \frac{f_x(\sqrt{y}) + f_x(-\sqrt{y})}{2\sqrt{y}} \tag{B.18}$$

B.3.3   Moments of Functions of Random Variables

If X is a random variable and $G(X)$ a function of this random variable then

$$\text{First moment: } \mathbf{mean} = \mathbb{E}[G(X)] = \int_{-\infty}^{+\infty} G(x)f_X(x)\mathrm{d}x \tag{B.19}$$

$$\text{Second moment: } \mathbf{variance} = \int_{-\infty}^{+\infty} (G(x) - \mathbb{E}[G(X)])^2 f_X(x)\mathrm{d}x \tag{B.20}$$

B.4 Jointly Distributed Random Variables

So far we have been only dealing with probability distributions of single random variables. However, we are often interested in probability statements concerning two or more random variables. In order to do this, we define the **joint cumulative distribution functions** of these random variables as

$$F_{XY}(x, y) = P[(X \leq x) \cap (Y \leq y)] \tag{B.21}$$

Properties:

- $F_{XY}(-\infty, -\infty) = 0$
- $F_{XY}(+\infty, +\infty) = 1$
- $F_{XY}(-\infty, y) = F_{XY}(x, -\infty) = 0.$

Joint Probability Density Function

$$f_{XY}(x, y) = \frac{\partial^2}{\partial x \partial y} F_{XY}(x, y)$$

$$\Leftrightarrow \quad F_{XY}(x, y) = \int_{-\infty}^{y} \int_{-\infty}^{x} f_{XY}(\xi_X, \xi_Y) d\xi_X d\xi_Y \tag{B.22}$$

The random variables X and Y are said to be statistically **independent** (the occurrence of one does not affect the probability of the other) **if and only if**

$$f_{XY}(x, y) = f_X(x) f_Y(y) \tag{B.23}$$

B.4.1 Moments of Jointly Distributed Random Variables

$$\mu'_{vv'} = \int_{-\infty}^{+\infty} \int_{-\infty}^{+\infty} x^v \cdot y^{v'} f_{XY}(x, y) dx dy \tag{B.24}$$

with $\mu'_{00} = 1$, $\mu'_{10} = \mathbb{E}[X] = \mu_X$, $\mu'_{01} = \mathbb{E}[Y] = \mu_Y$

Central Moments of Jointly Distributed Random Variables

$$\mu_{vv'} = \int_{-\infty}^{+\infty} \int_{-\infty}^{+\infty} (x - \mathbb{E}[X])^v (y - \mathbb{E}[Y])^{v'} f_{XY}(x, y) dx dy \tag{B.25}$$

with $\mu_{00} = 1$, $\mu_{10} = \mu_{01} = 0$, $\mu_{20} = \text{Var}[X] = \mu'_{20} - \mu'^2_{10} = \sqrt{\sigma_X}$, $\mu_{02} = \text{Var}[Y] = \mu'_{02} - \mu'^2_{01} = \sqrt{\sigma_Y}$

Covariance

$$
\begin{aligned}
\mu_{11} = K_{XY} &= \mathbb{E}[(X - \mu_X)(Y - \mu_Y)] \\
&= \int_{-\infty}^{y} \int_{-\infty}^{x} (\xi_Y - \mu_X)(\xi_Y - \mu_Y) f_{XY}(\xi_X, \xi_Y) d\xi_Y d\xi_Y \\
&= \mathbb{E}[XY] - \mu_X \mu_Y
\end{aligned}
\tag{B.26}
$$

If the random variables are uncorrelated then

$$
K_{XY} = 0 \Leftrightarrow \mathbb{E}[XY] = \mu_X \mu_Y
\tag{B.27}
$$

Normalized covariance (Coefficient of correlation)

$$
\rho_{XY} = \frac{K_{XY}}{\sigma_X \cdot \sigma_Y}
\tag{B.28}
$$

Properties:

$$
|\rho_{XY}| \leq 1 \Leftrightarrow |K_{XY}| \leq \sigma_X \sigma_Y
\tag{B.29}
$$

IMPORTANT NOTES:

- If $\rho_{XY} = \pm 1$ then the random variables X and Y are fully correlated, i.e., $Y = aX + b$
- If $\rho_{XY} = 0$ then the random variables X and Y are uncorrelated
- **IF** X,Y: Independent $\Rightarrow$ X,Y: Uncorrelated
- **IF** X,Y: Uncorrelated $\nRightarrow$ X,Y: Independent.

B.4.2   Binomial Distribution and Bernoulli Trials

Binomial distribution is a discrete probability distribution which expresses the number of success outcomes in a sequence of n independent experiments with boolean outcome 0 with probability p or $q = 1 - p$. A random variable following the binomial distribution with parameters $n \in \mathbb{N}$ and $p \in [0, 1]$, which gives the probability of getting exactly k successes in n trials given by

$$
B(k; n, p) = Pr[X = k] = \binom{n}{k} p^k (1 - p)^{(n-k)}
\tag{B.30}
$$

for $k = 0, 1, \ldots, n$, where

$$\binom{n}{k} = \frac{n!}{k!(n-k)!} \qquad (B.31)$$

is the so-called binomial coefficient. The formula has its roots in combinatorial analysis with k sucesses which occur with probability p^k and $n - k$ failures occur with probability $(1 - p)^{n-k}$. The k successes can occur anywhere among the n trial, and there are $\binom{n}{k}$ different ways of distributing them in this sequence. A single success/failure experiment ($n = 1$) is a special case of the binomial distribution called a Bernoulli trial.

B.5 Gaussian (Normal) Random Variables

A random variable X is considered to be normally distributed ($X \sim N(\mu, \sigma^2)$) if its probability density function is of the form:

$$f(x) = \frac{1}{\sqrt{2\pi}\sigma} \exp\left[-\frac{(x - \mu)^2}{2\sigma^2} \right] \qquad -\infty < x < +\infty \qquad (B.32)$$

where μ is its mean value and σ the standard deviation

If $\mu = 0$ and $\sigma = 1$ the random variable is the standard Gaussian random variable, denoted as $Z \sim N(0, 1)$ and the $\phi(x)$ is the standardized pdf. The corresponding probability distribution function $\Phi(x)$ is obtained by integration of $\phi(x)$ as

Fig. B.10 Perceptiles of the normal distribution for values less than one, two and three standard deviation away from the mean

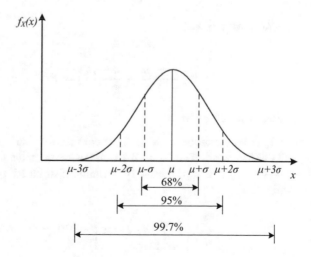

$$\Phi(x) = \int_{-\infty}^{x} \phi(\xi)d\xi$$

$$= \int_{-\infty}^{x} \frac{1}{\sqrt{2\pi}\sigma} \exp\left[-\frac{(\xi - \mu)^2}{2\sigma}\right]d\xi \qquad (B.33)$$

For the normal distribution, the values less than one standard deviation away from the mean account for 68.27% of the set, the values less than two standard deviations from the mean account for 95.45% and the values less than three standard deviations account for 99.73% (Fig. B.10).

Using the following change of variables

$$\zeta = \left(\frac{\xi - \mu}{\sigma}\right), \quad z = \left(\frac{x - \mu}{\sigma}\right) \qquad (B.34)$$

we obtain the cumulative distribution function of the **standard Gaussian** random variable Z

$$\Phi_Z(z) = \frac{1}{\sqrt{2\pi}\sigma} \int_{-\infty}^{z} e^{-\frac{\zeta^2}{2}} d\zeta$$

$$= \frac{1}{2} + \frac{1}{\sqrt{2\pi}\sigma} \int_{-\infty}^{z} e^{-\frac{\zeta^2}{2}} d\zeta \qquad (B.35)$$

Given a real number $x_a \in \mathbb{R}$ the probability that the random variable $X \sim N(\mu, \sigma^2)$ takes values less or equal x_a is given by

$$P(X \le x_a) = \begin{cases} \frac{1}{2} - \frac{1}{\sqrt{2\pi}\sigma} \int_{-\infty}^{\frac{x_a-\mu}{\sigma}} e^{-\frac{\zeta^2}{2}} d\zeta & \text{for } \mu \le x_a \\ \frac{1}{2} + \frac{1}{\sqrt{2\pi}\sigma} \int_{-\infty}^{\frac{x_a-\mu}{\sigma}} e^{-\frac{\zeta^2}{2}} d\zeta & \text{for } \mu \ge x_a \end{cases}$$

Generally we can write:

$$P(X \le x) = P\left(\frac{X - \mu}{\sigma} \le \frac{x - \mu}{\sigma}\right) = P(Z \le z) = \Phi\left(\frac{X - \mu}{\sigma}\right) \qquad (B.36)$$

Central Limit Theorem

Let X_n denote a sequence of independent random variables with mean value μ and finite variance σ^2. The Central Limit Theorem states that the arithmetic mean as $n \to +\infty$ will be approximately normally distributed, regardless of the underlying distribution,

$$\frac{1}{\sqrt{n}}\left[\left(\sum_{i=1}^{n} X_i\right) - \mu\right] \sim N(0, \sigma^2) \quad for \quad n \to \infty \qquad (B.37)$$

Because of its generality, this theorem is often used to simplify calculations involving finite sums of non-Gaussian random variables. However, attention is seldom paid to the convergence rate of the Central Limit Theorem. Consequently, the Central Limit Theorem, as a finite-sample distributional approximation, is only guaranteed to hold near the mode of the Gaussian, with huge numbers of observations needed to specify the tail behavior.

Properties of Gaussian Random Variables

1. The linear functions of Gaussian random variables remain Gaussian distributed.
   If $X \sim N(\mu_X, \sigma_X^2)$ then $Y = aX + b \sim N(\mu_Y = a\mu_X + b, \sigma_Y^2 = a^2\sigma_X^2)$

2. If X_1, X_2 are **independent** Gaussian random variables with mean values μ_1, μ_2 and standard deviations σ_1, σ_2, respectively, then the random variable $X = X_1 + X_2$ is also Gaussian with mean value $\mu = \mu_1 + \mu_2$ and standard deviation $\sigma = \sqrt{\sigma_1^2 + \sigma_2^2}$

3. The nth moment of the standard Gaussian random variable Z can be computed from:

$$E[Z^n] = \int_{-\infty}^{+\infty} z^n f_Z(z) \mathrm{d}z = \int_{-\infty}^{+\infty} z^n \frac{1}{\sqrt{2\pi}} \exp\left[-\frac{z^2}{2}\right] \mathrm{d}z \qquad (B.38)$$

4. A Gaussian random variable X with mean μ_X and standard deviation σ_X can be obtained from the standard Gaussian random variable:

$$X = \sigma_X Z + \mu_X \qquad (B.39)$$

Then the central moments of X are given from:

$$\mathbb{E}[(X - \mu_X)^n] = \mathbb{E}[(\sigma_X Z)^n] = \sigma_X^n \mathbb{E}[Z^n]. \qquad (B.40)$$

B.5.1 Jointly Distributed Gaussian Random Variables

Two random variables are jointly Gaussian distributed if the probability density function is given by

$$f_{XY}(x, y) = A exp\left\{-\frac{1}{2(1 - \rho_{XY}^2)}\left[\left(\frac{x - \mu_X}{\sigma_X}\right)^2\right.\right.$$
$$\left.\left. - 2\left(\frac{x - \mu_X}{\sigma_X}\right)\left(\frac{y - \mu_Y}{\sigma_Y}\right)\rho_{XY} + \left(\frac{y - \mu_Y}{\sigma_Y}\right)^2\right]\right\} \qquad (B.41)$$

where ρ_{XY} is the correlation coefficient of X and Y and

$$A = \frac{1}{2\pi\sigma_X\sigma_Y\sqrt{1 - \rho_{XY}^2}} \tag{B.42}$$

B.5.2   Gaussian Random Vectors

When working with all these random variables together, we will often find it convenient to put them in a vector $\mathbf{X} = [X_1, X_2, \ldots, X_n]^T$. We call the resulting vector a random vector. More formally, a random vector is a mapping from Ω to $\mathbb{R}^n$. It should be clear that random vectors are simply an alternative notation for dealing with n random variables, so the notions of joint pdf and cdf will apply to random vectors as well. Consider an arbitrary function $g(\mathbf{x}) : \mathbb{R}^n \to \mathbb{R}$ of the random vector $\mathbf{x}$. The mathematical expectation of this function can be defined as

$$\mathbb{E}[g(\mathbf{x})] = \int_{\mathbb{R}^n} g(x_1, x_2, \ldots, x_n) f_{X_1, X_2, \ldots, X_n}(x_1, x_2, \ldots, x_n) dx_1 dx_2 \ldots dx_n$$

$$= \int_{\mathbb{R}^n} g(\mathbf{x}) f_{\mathbf{X}}(\mathbf{x}) d\mathbf{x} \tag{B.43}$$

where $\int_{\mathbb{R}^n}$ is n consecutive integrations from $-\infty$ to ∞. If g is a function from $\mathbb{R}^n \to \mathbb{R}$, then the value of g is the element-wise expected values of the output vector. The covariance matrix $\mathbf{C}$ of the random vector is the $n \times n$ matrix whose entries are given by

$$C = \left[C_{ij}\right]_{n \times n} = \mathrm{Cov}[X_i, X_j] = \mathbb{E}\left[(X_i - \mu_{X_i})(X_j - \mu_{X_j})\right]. \tag{B.44}$$

The covariance matrix has a number of useful properties like being positive semi-definite and symmetric. The mean value vector $\boldsymbol{\mu}_\mathbf{x}$ of a random vector $\mathbf{x}$ is a vector containing the mean value of each component random variable

$$\boldsymbol{\mu}_\mathbf{x} = [\mu_{X_1}, \mu_{X_2}, \ldots, \mu_{X_n}]^T \tag{B.45}$$

The Multivariate Gaussian Distribution

One particularly important example of a probability distribution over random vectors is the multivariate Gaussian or multivariate normal distribution. A random vector $\mathbf{X} \in \mathbb{R}^n$ is said to have a multivariate normal (or Gaussian) distribution with mean $\boldsymbol{\mu} \in \mathbb{R}^n$ and covariance matrix $\mathbf{C}$ if its joint probability density is written as

$$f_{\mathbf{x}}(\mathbf{x}) = f_{X_1, X_2, \ldots, X_n}(x_1, x_2, \ldots, x_n; \boldsymbol{\mu}; C) = \frac{1}{(2\pi)^{n/2}|C|^{1/2}} exp\left(-\frac{1}{2}(\mathbf{x} - \boldsymbol{\mu})^T C^{-1}(\mathbf{x} - \boldsymbol{\mu})\right)$$

(B.46)

We note this random variable as $\mathbf{x} \sim N(\boldsymbol{\mu}_{\mathbf{x}}, C_{\mathbf{x}})$.

B.6 Transformation to the Standard Normal Space

Let $\mathbf{x}$ denote the realization of a vector of random variables with mean $\boldsymbol{\mu}_{\mathbf{x}}$ and covariance matrix $C_{\mathbf{x}}$. The objective is to transform $\mathbf{x}$ into a vector $\mathbf{y}$ of the same number of random variables with zero means and unit covariance matrix, i.e., $\mathbf{y}$ is a vector of uncorrelated and standardized random variables. In general, a second-moment transformation is written as

$$\mathbf{y} = \mathbf{a} + \mathbf{B}\mathbf{x}$$

(B.47)

were the vector $\mathbf{a}$ and the square matrix $\mathbf{B}$ contain unknown constants. Two equations with the unknowns $\mathbf{a}$ and $\mathbf{B}$ can be established by enforcing zero means and unit covariance matrix for $\mathbf{y}$:

$$\boldsymbol{\mu}_{\mathbf{y}} = \mathbf{a} + \mathbf{B}\boldsymbol{\mu}_{\mathbf{x}} = [0]$$

(B.48)

$$C_{\mathbf{y}} = \mathbf{B}C_{\mathbf{x}}\mathbf{B}^T = \mathbf{I}$$

(B.49)

$\mathbf{B}$ is the only unknown in Eq. (B.48). Multiplying now both sides of Eq. (B.49) with $\mathbf{B}^{-1}$ from the left and $\mathbf{B}^{-T}$ from the right yields the following expression for the covariance matrix of $C_{\mathbf{x}}$:

$$C_{\mathbf{x}} = \mathbf{B}^{-1}\mathbf{B}^{-T}$$

(B.50)

Hence, the unknown matrix $\mathbf{B}^{-1}$ is the **Cholesky decomposition** of $C_{\mathbf{x}}$, which can also be written as

$$C_{\mathbf{x}} = \mathbf{L}\mathbf{L}^T$$

(B.51)

where $\mathbf{L}$ is the lower triangular Cholesky decomposition of the covariance matrix, $\mathbf{B} = \mathbf{L}^{-1}$. Finally we obtain,

$$\mathbf{a} = \mathbf{L}^{-1}\boldsymbol{\mu}_{\mathbf{x}}$$

(B.52)

$$\mathbf{y} = \mathbf{L}^{-1}(\mathbf{x} - \boldsymbol{\mu}_{\mathbf{x}})$$

(B.53)

$$\mathbf{x} = \boldsymbol{\mu}_{\mathbf{x}} + \mathbf{L}\mathbf{y} \tag{B.54}$$

B.6.1   Nataf Transformation

Nataf transformation is a mathematical model for the transformation from original space to mutually independent (uncorrelated) standard normal one. The Nataf transformation does not require the joint pdf of all input random variables. However, the pdf of each random variable and the correlation matrix must be known. Consider the transformation

$$z_i = \Phi^{-1}(F_i(x_i)) \tag{B.55}$$

where the variables $\mathbf{z}$ are normally distributed with zero means and unit variances. However, they are correlated. Nataf assumption states that the random variables z_i are jointly normal. This is called the **Nataf** assumption. Under this assumption it can be shown that the correlation coefficient $\rho_{0,ij}$ between z_i and z_j is related to the correlation coefficient ρ_{ij} between x_i and x_j by the equation:

$$\rho_{ij} = \int_{-\infty}^{\infty} \int_{-\infty}^{\infty} \left(\frac{X_i - \mu_i}{\sigma_i} \right) \left(\frac{X_j - \mu_j}{\sigma_j} \right) \varphi_2(Z_i, Z_j, \rho_{0,ij}) \mathrm{d}Z_i \mathrm{d}Z_j \tag{B.56}$$

where φ_2 is the bivariate standard normal PDF:

$$\varphi_2(Z_i, Z_j, \rho_{0,ij}) = \frac{1}{2\pi\sqrt{1 - \rho_{0,ij}^2}} \exp\left\{ -\frac{Z_i^2 + Z_j^2 - 2\rho_{0,ij}Z_iZ_j}{2(1 - \rho_{0,ij}^2)} \right\} \tag{B.57}$$

Equation (B.56) can be solved iteratively to compute ρ_{ij}. This calculation is not trivial and a drawback of Nataf Transformation. Alternatively, the following approximate expression can be used for this calculation

$$\rho_{0,ij} = R_{ij}\rho_{ij} \tag{B.58}$$

where R_{ij} is approximated as polynomial:

$$R_{ij} = a + bV_i + cV_i^2 + d\rho_{ij} + e\rho_{ij}^2 + \tag{B.59}$$
$$+ f\rho_{ij}V_i + gV_j + hV_j^2 + krho_{ij}V_j + lV_iV_j$$

where V_i, V_j are the coefficient of variations for each random variable and the coefficients depend on the types of input variables. The expression of Eq. (B.58) provides with a reasonably small error (less than 1%) for all practical purposes. The Nataf model is valid under the conditions that the CDFs of x_i are strictly increasing and the correlation matrix of $\mathbf{x}$ and $\mathbf{z}$ be positive definite. It is an appealing transformation because it is invariant to the ordering of the random variables and a wide range of correlation values is acceptable. For $\mathbf{z}$ the covariance matrix is equal to the correlation matrix because the standard deviations are all zero. Hence, the Nataf transformation from $\mathbf{z}$ to standard uncorrelated space $\mathbf{y}$ is

$$\mathbf{z} = \mathbf{B}\mathbf{y} \tag{B.60}$$

where, similar to Eq. (B.51), $\mathbf{B}$ can be sought as the Cholesky decomposition

$$\boldsymbol{\rho}_0 = \mathbf{L}\mathbf{L}^T \tag{B.61}$$

where $\mathbf{L}$ is the lower triangular matrix obtained from the Cholesky decomposition of the correlation matrix $\boldsymbol{\rho}_0$ of $\mathbf{z}$. Thus, the following transformation stands:

$$\mathbf{y} = \mathbf{L}^{-1}\mathbf{z} \Leftrightarrow \mathbf{z} = \mathbf{L}\mathbf{y} \tag{B.62}$$

The transformation to standard normal space described in the previous section as well as the Nataf transformation are usually applied in the framework of FORM-SORM reliability analysis methods (see Chap. 4) which operate only in the standard normal space. In addition, they can be used for the generation of a set of correlated Gaussian or non-Gaussian (Nataf) random variables from a set of standard normal uncorrelated random variables following the inverse transformation:

$$x_i = F_i^{-1}(\Phi(z_i)) \tag{B.63}$$

B.7 Solved Numerical Examples

1. Consider a random variable X with pdf $f_X(x) = 1$ for $0 \leq x \leq 1$ and $f_X(x) = 0$ elsewhere. Let $Y = \sin(2\pi X)$ and $Z = \cos(2\pi X)$. Show that Y and Z are dependent and uncorrelated.

Solution:

It is obvious that: $Y^2 + Z^2 = 1$, therefore the random variables Y and Z are **statistically dependent** (related).

$$\mathbb{E}[Y] = \mathbb{E}[\sin(2\pi X)] = \int_0^1 \sin(2\pi x)1\mathrm{d}x = 0 \tag{B.64}$$

$$\mathbb{E}[Z] = \mathbb{E}[\cos(2\pi X)] = \int_0^1 \cos(2\pi x)1\mathrm{d}x = 0 \tag{B.65}$$

$$K_{YZ} = \mathbb{E}[Y - \mathbb{E}[Y]]\mathbb{E}[Z - \mathbb{E}[Z]] = \mathbb{E}[YZ]$$

$$= \int_{-\infty}^{+\infty} \sin(2\pi x)\cos(2\pi x) f_X dx$$

$$= \int_0^1 \frac{1}{2}\sin(4\pi x)dx = -\frac{1}{8\pi}[\cos(4\pi x)]_0^1 = 0 \qquad \text{(B.66)}$$

Although the random variables Y and Z are statistically dependent their covariance and thus their correlation coefficient is zero ($\rho = 0$). This means that the variables Y and Z are **uncorrellated**.

2. If the pdf of a random variable X is given by the following function

$$f_X(x) = \text{Prob}[x < X < x + dx] = \begin{cases} -\frac{3x^2}{4} + \frac{3x}{2} & \text{for } x \leq 2 \\ 0 & \text{otherwise} \end{cases}$$

find its expected value and its variance.

Solution:

$$\mu_X = \mathbb{E}[X] = \int_0^2 \left(-\frac{3x^3}{4} + \frac{3x^2}{2}\right) dx$$

$$= \left| -\frac{3x^4}{16} + \frac{x^3}{2} \right|_0^2 = 1.0$$

$$\sigma_X^2 = \mathbb{E}[X^2] - \mathbb{E}[X]^2 = \int_0^2 x^2 \left(-\frac{3x^2}{4} + \frac{3x}{2}\right) dx - (1.0)^2$$

$$= \left| -\frac{3x^5}{20} + \frac{x^4}{8} \right|_0^2 = 1.2 - 1 = 0.2$$

3. Suppose X is a continuous random variable with probability density $f_X(x)$. Find the distribution of $Y = X^2$.

Solution:

For $y \geq 0$

$$F_Y(y) = P[Y \leq y]$$

$$= P[X^2 \leq y]$$

$$= P[-\sqrt{y} \leq X \leq \sqrt{y}]$$

$$= F_X(\sqrt{y}) - F_X(-\sqrt{y})$$

The pdf can be found by differentiating $F_Y(y)$

$$f_Y(y) = \frac{1}{2\sqrt{y}}(F_X(\sqrt{y}) - F_X(-\sqrt{y}))$$

4. Consider the random variable X with pdf given by

$$f_X(x) = \begin{cases} 3x^2 & \text{for } 0 \le x \le 1 \\ 0 & \text{otherwise} \end{cases}$$

Find the pdf of the random variable $Y = 40(1 - X)$.

Solution:

Because the distribution function of Y looks at a region of the form $Y \le y$, we must first find that region of the x scale. Now

$$Y \le y \Rightarrow 40(1 - X) \le y$$
$$\Rightarrow X > 1 - \frac{y}{40}$$

Thus,

$$\begin{aligned} F_Y(y) &= P[Y \le y] \\ &= P[40(1 - X) \le y] \\ &= P\left[X > 1 - \frac{y}{40}\right] \\ &= \int_{1-y/40}^{1} f_X(x)\,dx \\ &= \int_{1-y/40}^{1} 3x^2\,dx \\ &= [x^3]_{1-y/40}^{1} \end{aligned}$$

or

$$F_Y(y) = \begin{cases} 0 & \text{for } y \le 0 \\ 1 - (1 - \frac{y}{40})^3 & \text{for } 0 \le y \le 40 \\ 1 & \text{for } y > 1 \end{cases}$$

So, the pdf $f_Y(y)$ can be found by differentiating the distribution function $F_Y(y)$ with respect to y

$$f_Y(y) = \begin{cases} \frac{1}{40}(1 - \frac{y}{40})^2 & \text{for } 0 \le y \le 40 \\ 0 & \text{otherwise} \end{cases}$$

5. Let X have the probability density function given by

$$f_X(x) = \begin{cases} 2x & \text{for } 0 \le x \le 1 \\ 0 & \text{otherwise} \end{cases}$$

Find the density function of $Y = -2X + 5$.

Solution:

Solving $Y = g(X) = -2X + 5$ for X, we obtain the inverse function

$$X = g^{-1}(Y) = \frac{5 - Y}{2} \tag{B.67}$$

g^{-1} is a continuous, one-to-one function from set $B = (3, 5)$ onto set $A = (0, 1)$. Since $\frac{dg^{-1}(y)}{dy} = -\frac{1}{2} \ne 0$ for any $y \in B$, then using Eq. (B.15) we get

$$f_Y(y) = f_X[g^{-1}(y)] \left| \frac{dg^{-1}(y)}{dy} \right| = 2(\frac{5-y}{2}) \left| -\frac{1}{2} \right| = \begin{cases} \frac{5-y}{2} & \text{for } 3 \le y \le 5 \\ 0 & \text{otherwise} \end{cases}$$

B.8 Exercises

1. Which of the following statements for two random variables X and Y are true?

    i. If X and Y are uncorrelated, they are also independent.
    ii. If X and Y are independent, $\mathbb{E}[XY] = 0$.
    iii. If X and Y are correlated, they are also dependent.

2. Let X be a continuous random variable with variance $\sigma_X^2 > 0$ and pdf $f_X(x) > 0$ symmetric around zero, i.e., $f_X(x) = f_X(-x), \forall x \in \mathbb{R}$. Let Y be a random variable given by $Y = aX^2 + bX + c$ with $a, b, c \in \mathbb{R}$. For which values of a, b and c are X and Y uncorrelated? For which values of $a, b,$ and c are X and Y independent?

3. Consider the random variable X with pdf given by

$$f_X(x) = \begin{cases} \frac{x+1}{2} & \text{for } -1 \le x \le 1 \\ 0 & \text{otherwise} \end{cases}$$

   Find the pdf of the random variables $Y_1 = X^2$ and $Y_2 = 1 - 2X$.

4. Let X be uniformly distributed over $(0, 1)$. Find the probability density function of the random variable $Y = e^X$.

5. Let X be a random variable with a probability density function given by

$$f_X(x) = \begin{cases} \frac{3}{2}x^2 & \text{for } -1 \le x \le 1 \\ 0 & \text{otherwise} \end{cases}$$

Find the density functions of the following variables: $U_1 = 3X$, $U_2 = 3 - X$ and $U_3 = X^3$.

Appendix C
Subset Simulation Aspects

C.1 Modified Metropolis- Hastings Algorithm

The modified Metropolis–Hastings (mMH) algorithm is a Markov chain Monte Carlo (MCMC) method for obtaining a sequence of random samples following some posterior distribution. Consider the vector $\mathbf{X} = [X_1, \ldots, X_n]$ containing n i.i.d random variables with pdf $f_X(\cdot)$. The modified version of the MH (mMH), which differs from the original Metropolis algorithm in the way the candidate state random variable is generated, utilizes a sequence of univariate proposal pdfs q_i (instead of a n-variate pdf of the original MH algorithm) in order to obtain the candidate state. The first step of mMH is to initialize the sample value for each random variable in $\mathbf{X}$ and then generate a sequence of random samples by computing $\mathbf{X}_k = [X_1^{(k)}, \ldots X_n^{(k)}]$ from $\mathbf{X}_{k-1} = [X_1^{(k-1)}, \ldots X_n^{(k-1)}]$ as follows:

- Generate a "candidate" sample $X_i^\star$ for sample i from the proposal distribution $q_i(X_i^{(k)}|X_i^{(k-1)})$.
- Find the acceptance probability

$$a = \min\left\{1, \frac{q_i(X_i^{(k-1)}|X_i^\star)f_X(X_i^\star)}{q_i(X_i^\star|X_i^{(k-1)})f_X(X_i^{(k-1)})}\right\} \tag{C.1}$$

- Generate samples u from $U[0, 1]$ and set

$$X_I^{(k)} = \begin{cases} X_I^\star, & \text{if } u < a \\ X_i^{(k-1)}, & \text{if } u \geq a \end{cases} \tag{C.2}$$

The proposal pdf $q(\cdot|\cdot)$ satisfies the symmetrical property ($q(A|B) = q(B|A)$) and thus, Eq. (C.1) can be written as

$$a = \min\left\{1, \frac{f_X(X_i^\star)}{f_X(X_i^{(k-1)})}\right\} \tag{C.3}$$

© Springer International Publishing AG 2018
V. Papadopoulos and D.G. Giovanis, *Stochastic Finite Element Methods*,
Mathematical Engineering, https://doi.org/10.1007/978-3-319-64528-5

Simulations show that the efficiency of mMH is insensitive to the type of the proposal pdf, and hence those which can be operated easily are the most preferable. For example, the uniform pdf $q_i(\cdot)$ centered at the current sample $X_i^{(k)}$ with width $2l_i$ is a good candidate, where l_i is the spread around the center. Parameter l_i is of crucial importance since it affects the size of the region covered by the Markov chain samples, and consequently it controls the efficiency of the method. Very small spreads of the proposal pdfs tend to increase the dependence between successive samples due to their proximity, thus slowing down convergence of the estimator and occasionally causing ergodicity problems. On the other hand, excessive large spreads may reduce the acceptance rate by increasing the number of repeated Markov Chain samples, thus slowing down convergence. The optimal choice for the spread of the proposal pdfs depends on a trade-off between acceptance rate and correlation due to proximity. Usually, the spread is chosen to be a fraction of the standard deviation of the starting samples, since no a priori knowledge is available that would lead to a convenient choice of spread.

In the framework of subset simulation described in Sect. 4.4.3, the samples generated using the mMH algorithm belong to the candidate vector $\mathbf{X}^{\star}$. In order to see whether to keep this or discard it, we check the location of $G(\mathbf{X}_k)$. If $G(\mathbf{X}^{\star}) \in G_i$ accept it as the next random vector, i.e., $\mathbf{X}_k = \mathbf{X}^{\star}$. Otherwise reject it and take the current vector as the next sample, i.e., $\mathbf{X}_k = \mathbf{X}_k$. This step ensures that the next sample always lies in G_i, so as to produce the correct conditioning in the samples.

Algorithm 1 Generate a random walk using modified Metropolis-Hastings algorithm

1: **procedure** GIVEN u_0, WIDTH
2:     $\tilde{u} = \text{rand}(\text{size}(u_0))$
3:     $\tilde{u} = u_0 + (\tilde{u} \times \text{width}) - \text{width}/2$
4:     $\text{ratio} = \dfrac{\exp(-0.5 \times \tilde{u}^2)}{\exp(-0.5 \times u_0^2)}$
5:     $\text{I} = \text{find}(\text{ratio} > 1)$
6:     $\text{ratio(I)} = 1$
7:     $d = \text{rand}(\text{size}(\tilde{u})) >= \text{ratio}$
8:     $\text{I} = \text{find}(d)$
9:     $\tilde{u}(I) = u_0(\text{I})$
10:     $g = \text{ones}(\text{size}(\tilde{u}))$
11:     $g(\text{I}) = 0$
12:     $\text{I}_{eval} = \text{find}(\text{any}(d, 1))$

C.2   SS Conditional Probability Estimators

The coefficient of variation (CoV) δ_1 of the probability of the first subset G_1, is directly calculated from Eq. (4.38) as

$$\delta_1 = \sqrt{\frac{1 - P_1}{P_1 N_1}} \qquad (C.4)$$

Since the Markov chains generated at each conditional level are started with samples selected from the previous simulation level and distributed as the corresponding target conditional pdf, the Markov chain samples used for computing the conditional probability estimators are identically distributed as the target conditional pdf. Thus, the conditional estimators P_i's are unbiased.

At the $(i - 1)$th conditional level, suppose that the number of Markov chains (seeds) is N_c and N_i/N_c samples have been simulated from each of these chains, so that the total number of Markov chain samples is N_i. Although the samples generated by different chains are in general dependent, it is assumed for simplicity that they are uncorrelated through the indicator function $I_{G_i}(\cdot)$. The covariance sequence $R_i(k)$ $i = 0, \ldots, (N/N_c - 1)$ can be estimated using the Markov chain samples at the (i)th conditional level by

$$R_i(k) = \frac{1}{N - kN_c} \sum_{j=1}^{N_c} \sum_{l=1}^{\frac{N}{N_c}-k} I_{jl}^i I_{j,l+k}^i - P_i^2 \tag{C.5}$$

The correlation coefficient at lag k of the stationary sequence is given by $\rho_i(k) = R_i(k)/R_i(0)$. Finally, since I_{jk}^i is a Bernoulli random variable $R_i(0) = P_i(1 - P_i)$ and so the variance of P_i is given by

$$\sigma_i^2 = \frac{P_i(1 - P_i)}{N}[1 + \gamma_i] \tag{C.6}$$

where

$$\gamma_i = \sum_{k=1}^{\frac{N}{N_c}-1} \left(1 - \frac{kN_c}{N}\right) \rho_i(k) \tag{C.7}$$

The CoV δ_i of P_i is thus given by:

$$\delta_i = \sqrt{\frac{1 - P_i}{P_i N}(1 + \gamma_i)} \tag{C.8}$$

The value of γ_i depends on the choice of the spread of the proposal pdfs. The CoV of P_f is upper bounded and the upper bound corresponds to the case that the conditional probability estimators are fully correlated. The actual CoV depends on the correlation between the P_i's. If all the P_i's were uncorrelated, then

$$\delta^2 = \sum_{i=1}^{M} \delta_i^2 \tag{C.9}$$

Although the P_i's are generally correlated, simulations show that δ^2 may well be approximated by Eq. (C.9).

References

Karhunen, K.: Über lineare Methoden in der Wahrscheinlichkeitsrechnung. Ann. Acad. Sci. Fennicae. Ser. A. I. Math.-Phys. **37**, 1–79 (1947)

Levy, P.: Processus stochastiques et mouvement Brownien. Gauthier-Villars (1965)

Loève, M.: Probability Theory, vol. II. Springer (1978). ISBN 0-387-90262-7

Mercer, J.: Functions of positive and negative type and their connection with the theory of integral equations. Philos. Trans. R. Soc. A **209**(441–458), 415–446 (1909)

Fredholm, E.I.: Sur une classe d'equations fonctionnelles. Acta Mathematica **27**, 365–390 (1903)

Ghanem, R., Spanos, P.: Stochastic Finite Elements: A Spectral Approach. Springer, Berlin (1991)

Elmendorf, D.A., Kriz, I., Mandell, M.A., May, J.P.: Rings, modules, and algebras in stable homotopy theory. Am. Math. Soc. Surv. Monogr. Am. Math. (1995)

Cameron, R.F., McKee, S.: Product integration methods for second-kind Abel integral equations. J. Comput. Appl. Math. **11**, 1–10 (1984)

Huang, S.P., Quek, S.T., Phoon, K.K.: Convergence study of the truncated Karhunen-?Loève expansion for simulation of stochastic processes. Int. J. Numer. Meth. Eng. **52**, 1029–1043 (2001)

Shinozuka, M., Deodatis, G.: Simulation of the stochastic process by spectral representation. Appl. Mech. Rev. ASME **44**(4), 29–53 (1991)

Papoulis, A.: Fourier Integral and Its Applications, 1st edn. McGraw-Hill, New York (1962)

Zerva, A.: Seismic ground motion simulations from a class of spatial variability models. Earthquaqe Eng. Struct. Dyn. **21**(4), 351–361 (1992)

Grigoriu, M.: Crossings of non-Gaussian translation processes. J. Eng. Mech. (ASCE) **110**(4), 610–620 (1984)

Grigoriu, M.: Simulation of stationary non-Gaussian translation processes. J. Eng. Mech. (ASCE) **124**(2), 121–126 (1984)

Zienkiewicz, O.C. CBE, FRS, Taylor, R.L., Zhu, J.Z.: The Finite Element Method Set, 6th edn. Its Basis and Fundamentals (1992). ISBN: 978-0-7506-6431-8

Brener, C.E.: Stochastic finite element methods-literature review. Internal Working Report No. 35791. University of Innsbruck, Austria (1991)

Hisada, T., Nakagiri, S.: Stochastic finite element method developed for structural safety and reliability. In: Proceedings of the 3rd International Conference on Structural Safety and Reliability. Trondheim, Norway, pp. 395–408 (1981)

Liu, W., Mani, A., Belytschko, T.: Finite element methods in probabilistic mechanics. Prob. Eng. Mech. **2**(4), 201–213 (1987)

Yamazaki, F., Shinozuka, M., Dasgupta, G.: Neumann expansion for stochastic finite element analysis. J. Eng. Mech. **114**(8), 1335–1354 (1988)

© Springer International Publishing AG 2018   135
V. Papadopoulos and D.G. Giovanis, *Stochastic Finite Element Methods*,
Mathematical Engineering, https://doi.org/10.1007/978-3-319-64528-5

Cameron, R.H., Martin, W.T.: The orthogonal development of non-linear functionals in series of Fourier–Hermite functionals. Ann. Math. **48**, 385–392 (1947)

Xiu, D., Karniadakis, G.E.: The Wiener-Askey polynomial chaos for stochastic differential equations. SIAM J. Sci. Comput. **24**, 619–644 (2002)

Papadrakakis, M., Papadopoulos, V.: Robust and efficient methods for the stochastic finite lement analysis using Monte Carlo simulation. Comp. Meth. Appl. Mech. Eng. **134**, 325–340 (1996)

Shinozuka, M., Deodatis, G.: Response variability of stochastic finite element systems. J. Eng. Mech. **114**, 499–519 (1991)

Deodatis, G.: Weighted integral method I: stochastic stiffness matrix. J. Eng. Mech. **117**(8), 1851–1864 (1991)

Matthies, H.G., Brenner, C.E., Bucher, C., Soares, G.: Uncertainties in probabilistic numerical analysis of structures and solids-Stochastic finite elements. Struct. Saf. **19**(3), 283–336 (1997)

Der Kiureghian, A., Ke, J.B.: The stochastic finite element method in structural reliability. Probab. Eng. Mech. **3**(2), 83–91 (1988)

Liu, W., Belytschko, T., Mani, A.: Random fields finite element. Int. J. Numer. Methods Eng. **23**, 1831–1845 (1986)

Brenner, C.E., Bucher, C.: A contribution to the SFE-based reliability assessment of non-linear structures under dynamic loading. Prob. Eng. Mech. **10**(4), 265–273 (1995)

Li, C.C., Der Kiureghian, A.: Optimal discretization of random fields. J. Eng. Mech. **119**(6), 1136–1154 (1993)

Vanmarcke, E.H., Grigoriu, M.: Stochastic finite element analysis of simple beams. J. Eng. Mech., ASCE **109**, 1203–1214 (1983)

Shinozuka, M.: Structural response variability. J. Eng. Mech. (ASCE) **113**, 825–842 (1987)

Papadopoulos, V., Kokkinos, O.: Variability response functions for stochastic systems under dynamic excitations. Probab. Eng. Mech. **28**, 176–184 (2012)

Teferra, K., Deodatis, G.: Variability response functions for beams with nonlinear constitutive laws. Probab. Eng. Mech. **29**, 139–148 (2012)

Deodatis, G., Miranda, M.: Generalized variability response functions for beam structures with stochastic parameters. J. Eng. Mech. **138**(9) (2012)

Miranda, M., Deodatis, G.: On the response variability of beams with large stochastic variations of system parameters. In: 10th International Conference on Structural Safety and Reliability, IASSAR, Osaka, Japan (2006)

Deodatis, G., Graham, L., Micaletti, R.: A hierarchy of upper bounds on the response of stochastic systems with large variation of their properties: random variable case. Probab. Eng. Mech. **18**(4), 349–364 (2013)

Arwade, S.R., Deodatis, G.: Variability response functions for effective material properties. Probab. Eng. Mech. **26**, 174–181 (2010)

Wall, F.J., Deodatis, G.: Variability response functions of stochastic plane stress/strain problems. J. Eng. Mech. **120**(9), 1963–1982 (1994)

Papadopoulos, V., Papadrakakis, M., Deodatis, G.: Analysis of mean response and response variability of stochastic finite element systems. Comput. Methods Appl. Mech. Eng. **195**(41–43), 5454–5471 (2006)

Papadopoulos, V., Deodatis, G., Papadrakakis, M.: Flexibility-based upper bounds on the response variability of simple beams. Comput. Methods Appl. Mech. Eng. **194**, 1385–1404 (2005)

Papadopoulos, V., Deodatis, G.: Response variability of stochastic frame structures using evolutionary field theory. Comput. Methods Appl. Mech. Eng. **195**(9–12), 1050–1074 (2006)

Kanai, K.: Semi-empirical formula for the seismic characteristics of the ground motion. Bull. Earthq. Res. Inst. Univ. Tokyo **35**(2), 308–325 (1957)

Tajimi, H.: A statistical method of determining the maximum response of a building structure during an earthquake. 2nd WCEE, Tokyo **2**, 781–798 (1960)

Clough, R.W., Penzien, J.: Dynamics of Structures. McGraw-Hill, New York (1975)

Rubinstein, R.Y.: Simulation and the Monte Carlo Method. Wiley, New York (1981)

Thoft-Chrinstensen, P.: Reliability and optimisation of structural systems. In: Proceedings of the 2nd IFIP WG7.5. Springer, London (1988)

Hasofer, A.M., Lind, N.C.: Exact and invariant second-moment code format. J. Eng. Mech. **100**, 111–121 (1974)

Papadrakakis, M., Tsobanakis, I., Hinton, E., Sienz, J.: Advanced solution methods in topology optimization and shape sensitivity analysis. In: Technical report, Institute of Structural Analysis and Seismic Research, NTUA, Greece (1995)

Madsen, H.O., Krenk, S., Lind, N.C.: Methods of Structural Safety. Prentice-Hall, Englewood Cliffs (1986)

Mckay, M., Beckman, R., Conover, W.: A comparison of three methods for selecting values of input variables in the analysis of output from a computer code. Technometrics **21**(2), 239–245 (1979)

Steinberg, H.A.: Generalized quota sampling. Nuc. Sci. Eng. **15**, 142–145 (1963)

Iman, R., Conover, W.: A distribution-free approach to inducing rank correlation among input variables. Commun. Stat. Simul. Comput. **11**(3), 311–334 (1982)

Metropolis, N., Rosenbluth, A.W., Rosenbluth, M.N., Teller, A.H., Teller, E.: Equation of state calculations by fast computing machines. J. Chem. Phys. **1**, 1087–1092 (1953)

Hastings, W.K.: Monte Carlo sampling methods using Markov chains and their applications. Biometrika **57**, 97–109 (1970)

Papadrakakis, M., Papadopoulos, V.: Lagaros N.D. Structural reliability analysis of elastic-plastic structures using neural networks and Monte Carlo simulation. Comp. Meth. Appl. Mech. Eng. **136**, 145–163 (1996)

Hurtado, J.E., Alvarez, D.A.: Neural-network-based reliability analysis: a comparative study. Comp. Meth. Appl. Mech. Eng. **191**(1–2), 113–132 (2002)

Lagaros, N.D., Papadrakakis, M.: Learning improvement of neural networks used in structural optimization. Adv. Eng. Softw. **35**, 9–25 (2004)

Papadopoulos, V., Giovanis, D.G., Lagaros, N.D., Papadrakakis, M.: Accelerated subset simulation with neural networks for reliability analysis. Comput. Methods Appl. Mech. Eng. **223–224**, 70–80 (2012)

Giovanis, D.G., Papadopoulos, V.: Spectral representation-based neural network assisted stochastic structural mechanics. Eng. Struct. **84**, 382–394 (2015)

Maltarollo, V.G., Honorio, K.M., Ferreira da Silva, A.B.: Applications of artificial neural networks in chemical problems. In: Artificial Neural Networks - Architectures and Applications (2013). https://doi.org/10.5772/51275

Mc Cutloch, W.S., Pitts, W.: A logical calculus of the Ideas imminent in nervous activity. Bull. Math. Biophys. **5**, 115–133 (1943)

Pitts, W., Mc Cutloch, W.S.: How we know universals: the perception of auditory and visual forms. Bull. Math. Biophys **9**, 127–147 (1947)

Hebb, D.O.: The Organization of Behavior. Wiley, New York (1949)

Hopfield, J.J.: Neural networks and physical systems with emergent collective computational abilities. Proc. Nat. Acad. Set. **79**, 2554, 2558 (1982)

Venn, J.: On the diagrammatic and mechanical representation of propositions and reasonings. Philos. Mag. Ser. 5, **10**(59) (1880)

de Moivre A.: De Mensura Sortis. Philos. Trans. **329** (1711)

Laplace, P.S.: Théorie analytique des probabilités. In: 3rd, in Euvres completes de Laplace, vol. 7 (1886)

Schroeder, L.: Buffon's needle problem: an exciting application of many mathematical concepts. Math. Teach. **67**(2), 183–186 (1974)

Sveshnikov, A.: Problems in Probability Theory, Mathematical Statistics and Theory of Random Functions. Dover, New York (1978)

Yaglom, A.M., Yaglom, I.M.: Challenging Mathematical Problems with Elementary Solutions. Dover, New York (1987)

Von, Mises R.: Probability, Statistics, and Truth (in German) (English translation, 1981: Dover Publications; 2nd Revised edn. (1939). ISBN 0486242145

Kolmogorov, A.N.: Grundbegriffe der Wahrscheinlichkeitrechnung, Ergebnisse Der Mathematik, (translated as Foundations of Probability, p. 1933. Chelsea Publishing Company, New York (1950)

Jaynes, E.T.: Probability Theory: The Logic of Science. Cambridge University Press, New York (2003)

Soong, T.T.: Fundamentals of Probability and Statistics for Engineers. Wiley, Chichester (2004)

Bertsekas, D.P., Tsitsiklis, J.N.: Introduction to Probability, 2nd edn. Athena Scientific, Belmont (2008)

Grigoriu, M.: Stochastic Calculus: Applications in Science and Engineering. Birkhäuser, Boston (2002)

Soong, T.T.: Random Differential Equations in Science and Engineering. Academic Press, New York (1973)

VanMarcke, E.: Random Fields: Analysis and Synthesis. The MIT Press, Cambridge (1988)

Papoulis, A.: Probability, Random Variables and Stochastic Processes, 3rd edn. McGraw-Hill, Auckland (1991)

Printed in the United States
By Bookmasters